Global Meat

Food, Health, and the Environment

Series Editor: Robert Gottlieb, Henry R. Luce Professor of Urban and Environmental Policy, Occidental College

For a complete list of books published in this series, please see the back of the book.

Global Meat

Social and Environmental Consequences of the Expanding Meat Industry

Edited by Bill Winders and Elizabeth Ransom

The MIT Press
Cambridge, Massachusetts
London, England

© 2019 Massachusetts Institute of Technology

All rights reserved. No part of this book may be reproduced in any form by any electronic or mechanical means (including photocopying, recording, or information storage and retrieval) without permission in writing from the publisher.

This book was set in ITC Stone Serif Std and ITC Stone Sans Std by Toppan Best-set Premedia Limited. Printed and bound in the United States of America.

Library of Congress Cataloging-in-Publication Data

Names: Winders, William, 1971- editor.
Title: Global meat : social and environmental consequences of the expanding meat industry / edited by Bill Winders and Elizabeth Ransom.
Description: Cambridge, MA : The MIT Press, [2019] | Series: Food, health, and the environment | Includes bibliographical references and index.
Identifiers: LCCN 2019001208 | ISBN 9780262537735 (paperback : alk. paper)
Subjects: LCSH: Meat industry and trade--Environmental aspects. | Meat industry and trade--Social aspects.
Classification: LCC HD9410.5 .G56 2019 | DDC 338.4/76649--dc23 LC record available at https://lccn.loc.gov/2019001208

10 9 8 7 6 5 4 3 2 1

For Violet and Sam.—BW

For my family.—ER

Contents

Series Foreword ix
Preface xi

1 Introduction to the Global Meat Industry: Expanding Production, Consumption, and Trade 1
Bill Winders and Elizabeth Ransom

I Global Forces 25

2 Corporate Concentration in Global Meat Processing: The Role of Feed and Finance Subsidies 31
Philip H. Howard

3 Aquatic CAFOs: Aquaculture and the Future of Seafood Production 55
Conner Bailey and Nhuong Tran

II From Global to Local 75

4 China's Global Meat Industry: The World-Shaking Power of Industrializing Pigs and Pork in China's Reform Era 79
Mindi Schneider

5 Amerindians, Mestizos, and Cows in the Ecuadorian Amazon: The Silvopastoral Ecology of Small-Scale, Sustainable Cattle Ranching 101
Thomas K. Rudel

6   Cheap Meat and Cheap Work in the U.S. Poultry Industry: Race, Gender, and Immigration in Corporate Strategies to Shape Labor   121
    Carrie Freshour

III   Consequences and Considerations   141

7   Contributions to Global Climate Change: A Cross-National Analysis of Greenhouse Gas Emissions from Meat Production   145
    Riva C. H. Denny

8   Livestock Intensification Strategies in Rwanda: Ethical Implications for Animals and a Consideration of Potential Alternatives   167
    Robert M. Chiles and Celize Christy

9   Conclusions about the Global Meat Industry: Consequences and Solutions   185
    Elizabeth Ransom and Bill Winders

References   201
List of Contributors   235
Index   239

Series Foreword

Global Meat: Social and Environmental Consequences of the Expanding Meat Industry is the sixteenth book in the Food, Health, and the Environment series. The series explores the global and local dimensions of food systems and the issues of access, social, environmental, and food justice, and community well-being. Books in the series focus on how and where food is grown, manufactured, distributed, sold, and consumed. They address questions of power and control, social movements and organizing strategies, and the health, environmental, social, and economic factors embedded in food-system choices and outcomes. As this book demonstrates, the focus is not only on food security and well-being but also on economic, political, and cultural factors and regional, state, national, and international policy decisions. Food, Health, and the Environment books therefore provide a window onto the public debates, alternative and existing discourses, and multidisciplinary perspectives that have made food systems and their connections to health and the environment critically important subjects of study and for social and policy change.

Robert Gottlieb, Occidental College
Series Editor (gottlieb@oxy.edu)

Preface

Each of us came to this project out of personal concern and scholarly interest. We both have concerns, shared by many others, about how the meat industry has transformed over the past half-century, as it has become increasingly industrial in nature. Confined animal feeding operations (CAFOs) are a central example of such techniques, bringing thousands of animals into small, confined, unnatural spaces. The growth of CAFOs has facilitated the sharp increase in the production of animals—cows, pigs, chickens, and fish—that are slaughtered to be consumed as meat. As the number of animals has increased, the pace of slaughtering terrestrial animals has likewise quickened, threatening the health of slaughterhouse workers who struggle to keep up with the faster production line speeds. We are therefore concerned with how this transformation has affected workers, farmers, peasants, consumers, and animals. In different ways, there are important reasons to be concerned for the health and well-being of these various groups, many of whom have suffered in increasingly violent ways at the hands of the meat industry. The changes in the meat industry have also threatened the environment by contributing to global climate change and by polluting local water sources, air, and land. This transformation of the meat industry even poses economic threats to farmers and smallholders (small farmers with limited land, labor, and capital), whose land and livelihoods are often at great risk. There is much to be concerned about.

This personal concern led to our scholarly interest in trying to understand the global dynamics of the meat industry over the past several decades. Our project began with a coauthored paper that aimed to outline the global patterns of meat production, consumption, and trade. We presented this paper at the Annual Meeting of the Rural Sociological Society in the summer of 2016, and the response was terrific. Not only did the paper draw much interest and many questions and comments, but also several scholars spoke with us after the presentation and offered to participate in the project. Analyzing the contours of the global meat industry was a task that many scholars wanted to join. As a result of this response, we decided to host a scholarly workshop as a first step toward an edited volume that would draw together the insights of several scholars. The two-day workshop was hosted by Georgia Tech, and it was supported by several partners around Georgia Tech's campus: the School of History and Sociology, the Brook Byers Institute for Sustainable Systems, and the SLS (Serve-Learn-Sustain) Fellows Program in Food, Energy, and Water Systems (FEWS). All of the authors in this volume participated in this workshop, presenting papers that would ultimately become the chapters in this book. We are grateful to the campus partners at Georgia Tech who helped us take this important first step toward this edited volume.

Edited volumes such as this one are different from monographs not only in the many voices and perspectives that they bring together but also because of the wealth of knowledge that lies behind them. The group of authors behind these chapters know, as a group, much more than either of us knew individually about the global meat industry. So this has been an important opportunity for us to better understand the global meat industry from different vantage points, and we are grateful for that. Edited volumes can be a bit trickier than monographs in the publishing world. Consequently, we appreciate Beth Clevenger, our editor at MIT Press, and her support and enthusiasm for this book all the more. Beth has also given us important guidance and invaluable comments and suggestions along the way. We are grateful for her support for this project and her patience with us as we worked to finish it. We thank Kathleen Caruso in her work on this book as Senior Editor at MIT Press, and we also appreciate Julia Collins for her tremendous work as the copy editor for this volume. We would also like to thank the three anonymous reviewers of this volume. Their feedback on an earlier draft of this work was invaluable. Finally, thanks to the coauthors of this volume who spent considerable time and energy on their chapters, including several iterations of revisions to ensure a coherent manuscript. Of course, any mistakes are attributable only to us.

1 Introduction to the Global Meat Industry: Expanding Production, Consumption, and Trade

Bill Winders and Elizabeth Ransom

Over the past five decades, the global meat industry has emerged as one of the most important factors shaping our lives and the world around us. Yet, the expanse of this industry can be difficult to see precisely because of its wide reach. The production and consumption of meat has reached such a global scale that a hamburger can link together people and businesses across the globe: genetically engineered soybeans from U.S.-based Monsanto, a soybean farmer in Brazil, a cattle farmer and a slaughterhouse worker in Australia, and a consumer at McDonald's in Japan. The expanding reach and interconnection of this industry across the globe has been driven by new trade policies that have encouraged a growth in global trade, government subsidization of specific crops, and corporations, including ADM, Cargill, Monsanto, JBS, WH Group Limited, and Tyson Foods Inc. This growth in the meat industry has brought with it several potential consequences that threaten to worsen a variety of social and environmental issues already seen as problematic: climate change, clean water supplies, food insecurity and world hunger, consumers' health, workers' rights and well-being, and the treatment of animals. Consequently, understanding the new realities of the global meat industry is of paramount importance.

Many questions arise in grappling with the global expansion in the meat industry. What forces have driven the expansion of the global meat industry? What role have corporations played, and how powerful are these businesses? Have national governments supported or limited the expansion in the meat industry? How has this global expansion contributed to the adoption of industrial production methods in raising animals—chickens, cows, pigs, and even fish—and what have been the consequences for the environment and the animals themselves? What has the increased size and global reach of corporations meant for the working conditions of slaughterhouse workers and the independence of farmers? What opportunities exist for production that is more sustainable for the environment and that recognizes the important ethical concerns related to animals and workers?

The chapters that follow in this book address these questions by drawing on concrete examples. This book brings together scholars with expertise in different dimensions of the complex web of the global meat industry: corporate concentration and power; industrial production of animals and fish; meat production in Africa, Asia, and North and South America; greenhouse gas (GHG) emissions from livestock; environmentally sustainable production by small-scale farmers; and the experiences of workers in processing plants. While focusing on this range of issues, three themes cut across the chapters. First, governments and corporations have important roles in shaping the structure of the global meat industry. Second, increasing meat consumption for a growing global population can create a fundamental contradiction: despite rising meat production and consumption, there is a threat to food security for many smallholders and other people around the world. And third, the global meat industry contributes to social and environmental injustice as it relates to people, land/territory, and animals. The chapters in this book help to highlight these issues in the global meat industry.

First, governments and corporations have each played a central role in the expansion of the global meat industry. The value of this global meat production has increased from about $65 billion (in constant 2004–2006 US$) in 1961 to $366 billion in 2014—an increase of more than 500 percent.[1] A handful of corporations have come to dominate the meat industry as it expanded over the past five decades. Today, three corporations are particularly noteworthy: JBS (Brazil), WH Group (China), and Tyson (United States). Each of these corporations has received substantial subsidies from their respective governments (see chapter 2). The wide reach of such corporations, with the help of governments, has undermined the potential for democratic or local control over food supply.

Second, the expansion of the global meat industry has reflected, though clearly outpaced, the planet's increasing human population. From 1960 to 2016, the world's population increased from 3 billion people to 7.4 billion people, an increase of almost 150 percent. At the same time, the global production of meat during the same period increased from about 45 million metric tons (MMT) to 259 MMT, or an increase of more than 500 percent. As the population increased, annual per capita meat consumption doubled from 20 kg per year in 1961 to 40 kg per year. Thus, the global meat industry has provided one possible avenue for feeding the world's growing population. However, we demonstrate in this volume that this solution has undermined food security because of the corporate control and centralized economic power in the meat industry as well as in the concomitant expansion of feed grain production (e.g., Turzi 2017; Winders 2017).

Third, the global expansion of the meat industry has contributed to several problems, including more dangerous conditions for workers, environmental degradation, and harmful treatment of animals. In its expansion, the meat industry has adopted production techniques that have increasingly put workers at risk in terms of their health and their economic well-being. The industry has also developed intensive production models that rely on housing and raising animals in more concentrated settings, creating stressful and unhealthy conditions for animals and neighboring communities. These intensive methods also pose threats to the environment as water usage increases, on the one hand, and water sources become threatened by pollution from animal waste, on the other hand. The increasing number of animals in the meat industry has also meant the need for more land for raising and housing animals, grazing some animals, and feed grain production. Thus, this expanding industry has also threatened the access to land for small farmers and peasants across the global (e.g., Lapegna 2016). Such production techniques have been adopted and spread across the globe—from the United States to Brazil to China—thereby prompting the proliferation of these problems for workers, smallholders, animals, and the environment.

Before turning to the chapters and the discussions of these issues, let us put into perspective what we mean when we say "the global meat industry has changed over the past several decades." This will provide the backdrop for the chapters that follow. Specifically, we will outline the global patterns in meat production, consumption, and trade (exports/imports). More importantly, we will consider what such trends tell us about the scope and impact of the meat industry today.

The Global Meat Industry, 1960–2016

Several dimensions of the global meat industry are important to consider: production, consumption, economic value, animals consumed, and corporate concentration, among others. We will look briefly at these dimensions since they reveal how much the industry has grown, expanded and shifted geographically, and gained in its economic significance to the broader world economy. While some scholars have recently examined changes in the global diet and the emergence of the "neoliberal diet" (Otero 2018), this research has tended to focus on the increased global consumption of oils, sugars, and processed foods. Although meat has not received as much attention in this evolving global diet, rising meat consumption should undoubtedly be part of discussions of the changing global diet (Weis 2013a). This is clear when we look at how these dimensions of the global meat industry—consumption, production, and total value—have changed over the past five decades.

Global meat consumption and production have risen sharply and steadily since 1960. Turning to consumption first, figure 1.1 shows the growth in global per capita meat consumption from 1961 to 2013. Annual global per capita meat consumption was 20 kg in 1961, and it climbed steadily over the next five decades. By 1988, per capita meat consumption had risen by 50 percent to 30 kg/year. And in 2013, it reached 40 kg/year—double the per capita meat consumption in 1961. This increase in global consumption was centered on pigs and chickens, as per capita beef consumption actually declined during this period from 11.5 kg/capita in 1976 to 9.3 kg/capita in 2013. Per capita consumption of chickens increased the most, from 4.8 kg in 1976 to 15 kg in 2013. During the same period, the per capita consumption of pigs increased from 10.2 kg to 16 kg.

The increase in per capita meat consumption, of course, has not been spread evenly across the globe. Some countries and regions saw larger increases than others, and some even saw a decrease in meat consumption. Thinking about the geography of global meat consumption leads to several questions. Is meat consumption concentrated in particular areas? Where has meat consumption increased the most over the past several decades? Has meat consumption increased only the global north? Countries in the global north—including Australia, Canada, the European Union, and the United States—do have high levels of per capita meat consumption. At

Introduction to the Global Meat Industry

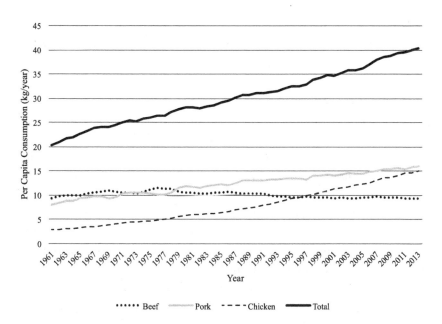

Figure 1.1
Global per capita meat consumption, 1961–2013
Source: FAO 2019.

113.9 kg/capita, the United States has among the highest per capita total meat consumption. In recent years, some other countries have also had a per capita meat consumption increase. China, Japan, and South Korea all saw dramatic increases in per capita meat consumption, from 5 kg/capita in 1961 to as much as 68 kg/capita in 2013. Even in countries that the FAO identifies as the "Least Developed Countries," annual per capita meat consumption increased from 6.4 kg in 1961 to 10.9 in 2013, with the most significant increase occurring between 1994 and 2010, from 6.6 kg to 10.4 kg. Therefore, the total increase in per capita meat consumption has been smaller for the poorest countries than in the global north, but it increased nevertheless.

Contrary to the overall trends, some wealthier countries experienced a decline in per capita meat consumption. Most notably, total per capita meat consumption in some European countries has decreased since the 1990s. For example, per capita meat consumption in Denmark decreased from 99.3 kg in 1995 to 80.1 kg in 2013.[2] France saw its per capita meat consumption decline from a high of 91.8 kg in 1998 to 79.8 kg in 2013. The Netherlands also saw its per capita meat consumption fall from 92.8 kg in 1998 to 77.9 in 2013. Even in the United States, per capita meat consumption has declined recently from a high of 124.2 kg in 2006 to 113.9 in 2013. Other countries had stable per capita meat consumption. In Germany, for example, meat consumption was 82.5 kg in 1998 and 82.7 in 2013. Despite these examples of declining or stable per capita meat consumption in the global north, the overall global trend has been an increase in per capita meat consumption.

Furthermore, at the same time that global per capita meat consumption increased, the world's population grew from 3 billion people in 1960 to 7.4 billion people in 2016. Therefore, per capita meat consumption increased as the size of the population also increased. While per capita meat consumption increased by about 100 percent from 1961 to 2013, the world's population increased by about 146 percent. Thus, the increase in total (i.e., not per capita) global meat consumption was tremendous, as the world came to have more than twice as many people who, on average, consume twice as much meat. How we have produced enough meat for such an explosion in consumption can be better understood by looking at the trends in global meat production.

Figure 1.2 shows that total global meat production of land animals—cows, pigs, and chickens—increased from about 45 million metric tons in 1960 to 259 MMT in 2016. While per capita meat consumption increased by 100 percent and the world's population increased by 146 percent during this period, global meat production increased by almost 500 percent. During this period, beef production increased from 23 MMT to 60 MMT, pork production increased from 19 MMT to 109 MMT, and chicken production increased from about 2 MMT to 88 MMT. Thus, while production increased for each type of meat, the most significant increases came in the production of pork and chicken. Another way to view this increase in global meat production is by looking at the number of animals slaughtered each year, as seen in figure 1.3. In 1961, about 173 million cows were slaughtered, 376 million pigs, and 6.5 billion chickens. In 2014, the slaughter of animals for meat rose to 300 million cows, 1.4 billion pigs, and 62 billion chickens. Here again, the slaughter of cows was the smallest increase, though it was still a 73 percent increase. The really significant increases, of course, came in the numbers of pigs and chickens slaughtered each year:

Introduction to the Global Meat Industry

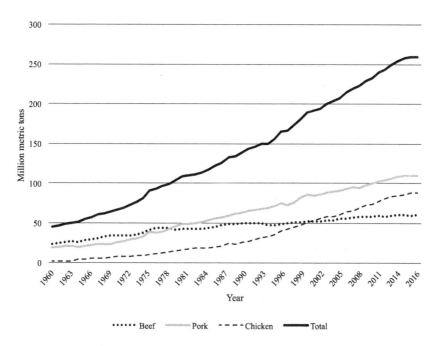

Figure 1.2
Global meat production, 1960–2016
Source: USDA FAS n.d.

from 1961 to 2014, the number of pigs slaughtered for meat increased by about 270 percent, and the numbers of chickens killed saw an increase of more than 900 percent. The expanding global meat industry, then, rested on an incredible increase in the number of animals kept and slaughtered for meat. This expansion in meat production, as this volume aims to document and explain, required a tremendous amount of resources—land, feed grains, labor, water, and money—and it brought with it a series of substantial consequences—environmental pollution, animal and human exploitation, and even food insecurity. This expansion of global meat production, which has relied on raising and killing so many more animals over the past several decades, has rested on the spread of concentrated feeding operations (CAFOs). These industrial livestock operations concentrate large numbers of animals in relatively small geographic spaces to maximize efficiency, often at the expense of the health and general well-being of the animals, the environment, and surrounding communities.

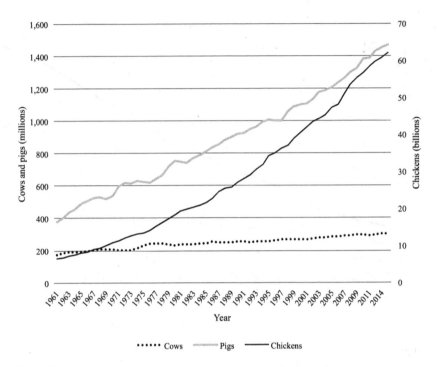

Figure 1.3
Number of animals slaughtered in global meat production, 1961–2014
Source: FAO 2019.

Given the sharp increases in consumption and production, it should not be surprising that the economic value created by the global meat industry has likewise increased dramatically, as seen in figure 1.4. In 1961, the total economic value produced by global meat production was about US$65 billion (in constant 2004–2006 dollars). By 1990, the economic value from meat production was US$176 billion, and it was US$366 billion by 2014. Since 1990, the economic value of global meat production more than doubled, controlling for inflation. This economic value is not spread at all evenly across farmers, peasants, workers, and corporations, or between countries in the global south and north. Rather, corporations in the global north (especially, in the United States) and industrializing countries (specifically, Brazil and China) have reaped the lion's share of this greater value in meat production.

Introduction to the Global Meat Industry

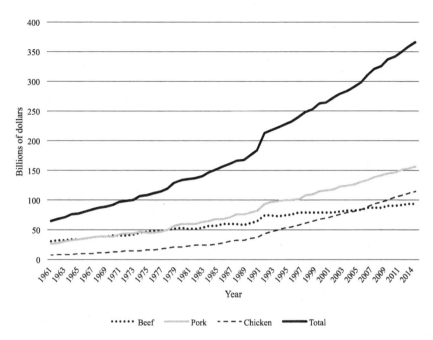

Figure 1.4
Value of global meat production, 1961–2014 (constant 2004–2006 US$)
Source: FAO 2019.

The global meat industry now consumes more animals, which also means using more feed grains, more land, and more water. As it has expanded production, the economic value of the global meat industry has grown, more than doubling just since 1990. All the while, the geographic reach of the meat industry has expanded to encompass the globe as never before. This geographic expansion, of course, has depended on the increased international trade of meat.

Trade in the Global Meat Industry

In addition to production, consumption, and value, taking a close look at the dynamics of international trade also helps to illuminate the contours of the global meat industry. Just as global meat production and consumption have increased, so too has international trade in meat. From 1960 to 2015, the meat exports increased from 2.6 MMT to 27 MMT. And since the 1990s, the global meat trade has expanded most rapidly, more than doubling from 12.6 MMT to 27 MMT in just eighteen years, from 1998 to 2015. Similar amounts of each type of meat are exported: in 2015, 10 MMT of beef and chicken along with 7 MMT of pork. Given the total production of each type of meat (see figure 1.2), beef is exported at the highest rate (16 percent of overall beef production in 2015), then chicken (11.6 percent), followed by pork (6.5 percent). As with production and consumption, of course, there is a geographic component to trade in the global meat industry. Which countries are the primary exporters and importers in the world?

A substantial portion of meat exports comes from countries in the global north. And geographically, the global meat industry is anchored by exports from North and South America. Table 1.1 shows the world's leading exporter of beef, pork, and chicken in 2015. Among these leading meat-exporting countries, Brazil and the United States stand out as the top exporters of each type of meat: Brazil was first in chicken, third in beef, and

Table 1.1

Top five countries for global meat exports, 2015

	Country	Total exports (MMT)	Percent of world exports	Total market share (percent)
Beef	Australia	1.854	19.4	73.6
	India	1.806	18.9	
	Brazil	1.705	17.8	
	United States	1.028	10.8	
	New Zealand	0.639	6.7	
Pork	EU	2.388	33.1	93.2
	United States	2.241	31.1	
	Canada	1.236	17.1	
	Brazil	0.627	8.7	
	China	0.231	3.2	
Chicken	Brazil	3.841	37.4	86.6
	United States	2.866	27.9	
	EU	1.177	11.5	
	Thailand	0.622	6.1	
	China	0.401	3.9	

Source: Calculated using data from USDA FAS n.d.

fourth in pork; and the United States is second in pork and chicken, and fourth in beef. In addition, it is also noteworthy that two exporters dominate the markets in pork and chicken—the European Union and United States in pork, and Brazil and the United States in chicken—while the beef export market is more evenly divided among three or four exporters. India stands out among beef exporters as the world's second-leading exporter of beef in 2015.[3] The countries accounting for at least 10 percent of exports of meat include Australia, Brazil, Canada, the EU, India, and the United States. Therefore, we can see that much exported meat comes from countries in the global north, and the Americas figure prominently in this export.

The pattern of global meat imports is different from that of exports, with countries in Asia and emerging industrial countries accounting for a substantial portion of global meat imports. Table 1.2 shows the countries that imported the most beef, pork, and chicken. Among these leading meat-importing countries, Japan is the most notable. In 2015, Japan was

Table 1.2

Top five countries for global meat imports, 2015

	Country	Total imports (MMT)	Percent of world imports	Total market share (percent)
Beef	United States	1.529	20.2	51.9
	Japan	0.707	9.3	
	China	0.663	8.7	
	Russia	0.625	8.2	
	South Korea	0.414	5.5	
Pork	Japan	1.270	19.0	65.6
	China	1.029	15.4	
	Mexico	0.981	14.7	
	South Korea	0.599	9.0	
	United States	0.504	7.5	
Chicken	Japan	0.936	10.8	46.4
	Saudi Arabia	0.930	10.7	
	Mexico	0.790	9.1	
	EU	0.728	8.4	
	Iraq	0.640	7.4	

Source: Calculated using data from USDA FAS n.d.

the world's top importer of pork and chicken, and it was the second-biggest importer of beef, behind the United States. In addition to Japan and the United States, the other top beef-importing countries were China, Russia, and South Korea. The top pork-importing countries behind Japan were China, Mexico, South Korea, and the United States. And, the top chicken-importing countries behind Japan were Saudi Arabia, South Korea, the EU, and Iraq. We can therefore see two important patterns in global meat imports. First, East Asia—Japan, South Korea, and China—is the region that imports the most meat, particularly beef and pork, though Japan is also the leading importer of chicken. Second, emerging "middle class" and industrial countries—China, Mexico, Russia, and South Korea—are among the leading importing countries. These countries have been rising in the world economy during the past five decades, and their national diets have shifted to include more meat as they have risen. Annual per capita meat consumption has increased dramatically in each of these countries from 1961 to 2013: from 3.5 kg to 57.6 kg in China, 22.5 kg to 60.7 kg in Mexico, 33.5 kg to 70.1 kg in Russia, and 3.9 kg to 63.3 kg in South Korea.[4] Japan saw a similar increase in its annual per capita meat consumption, from 5.1 kg in 1961 to 49.1 kg in 2013. These sharply increasing levels of meat consumption rested on the expansion of meat imports through the global meat industry.

These two tables reveal a couple of important points about the shape of the current meat trade. First, there is a geographical dimension, in which much meat is exported from North and South America and imported by Asia. There are, of course, other exporters and importers. India is a leading exporter of beef, and its main markets are regional (e.g., Thailand and Vietnam among others). Nevertheless, about a quarter of the world's beef exports, more than half of its pork exports, and almost two-thirds of its chicken exports come from North and South America, especially the United States and Brazil. Second, the global meat trade is uneven in terms of the power of exporting states versus importing states. Many exporting states are nations in the global north—Australia, Canada, the EU, New Zealand, and the United States—which means that they have disproportionate power in the interstate system relative to meat-importing countries that tend to be middle-income countries (e.g., Mexico and South Korea) or, especially, the global south (e.g., Iraq). Of particular importance, countries in the global north tend to have more influence over the terms of trade, such as the level of tariffs and trade barriers (Chase-Dunn, Kawano, and Brewer 2000). This imbalance in power in trade is reinforced by the different levels of market concentration among exporting countries and importing countries. Among the top exporting countries, the top five exporters accounted for 73 percent of beef exports, 86 percent of chicken exports, and 93 percent of pork exports in 2015. By contrast, the top five importers accounted for substantially smaller portions of total world imports: about 52 percent of beef, 46 percent of chicken, and 66 percent of pork. This tends to give exporting nations more leverage in the market (Howard 2016). All of this contributes to unequal exchange between richer and poorer countries.

Factors behind the Expanding Global Meat Industry

Given the growth in meat consumption and production around the globe, as well as the increased international meat trade, we should take a moment to consider some of the factors that have contributed to this global change in diets. While there are many factors that have played a role in the expanding global meat industry, three are especially notable: an increase in global feed grain production, a global shift toward market-oriented policies and liberalization, and increased corporate concentration. As the following discussion demonstrates, these three factors are interrelated.

First, one important factor contributing to this expansion in the global meat industry has been an increase in the production of feed grains across the globe. World production of corn (maize) and soybeans, in particular, has increased significantly: from 1990 to 2015, corn production doubled from 481 MMT to 967 MMT, and soybean production more than tripled from 104 MMT to 319 MMT (Winders 2017, 3, 11). In 2015, the global production of these feed grains—a combined total of 1,286 MMT—was larger than the production of the two major food grains, wheat and rice, which were 735 MMT and 430 MMT, respectively, for a combined total of 1,165 MMT.[5] And the amount of land devoted to these food grains and feed grains were as follows: 224 million hectares of wheat, 176 million hectares of corn, 158 million hectares of rice, and 120 million hectares of soybeans. This global expansion of feed grain production was fueled partly by national policies that offered support in the form of farm income subsidies and in the expansion of land. Furthermore, advances in agricultural technology helped to fuel this expansion of world food grain production: in 1996, genetically engineered (GE) corn and soybeans became available. Today, most of the soybeans produced are GE soybeans, and most GE feed grain production occurs in North and South America, with the United States and Brazil being the world's two largest producers and exporters of corn and soybeans. Together these two countries account for about 80 percent of world soybean exports and about 60 percent of world corn exports. This increased feed grains production made possible the expansion of the global meat industry.

Second, a shift in policies toward more liberal and market-oriented policies contributed to the expansion of the global meat industry. The creation of the World Trade Organization (WTO) in 1995 and the spread of bilateral and regional free trade agreements from the 1990s onward were at the heart of liberalization and the expansion of international trade. This is especially true for agriculture, since prior to the WTO formation, countries had largely protected their agricultural sectors from free trade. The numerous regional free trade agreements in the 1990s included the North American Free Trade Agreement (NAFTA), the Free Trade Area of the Americas, the European Economic Area, Mercosur (founded by Argentina, Brazil, Paraguay, and Uruguay), and the Asia-Pacific Economic Cooperation forum. At the same time as these regional trade agreements were forming, the number of bilateral free trade agreements increased dramatically. Regional associations, such as the European Free Trade Association, also made an increasing number of agreements with individual countries. In addition to the spread of free trade agreements, communism fell in Eastern Europe and the Soviet Union, and markets in China expanded. This decline of communism opened markets for feed grains and meat that had previously been extensively regulated or even closed off.[6] Such policies, agreements, and national changes allowed for greater trade both in meat and in feed grains, which contributed to various nations increasing their own meat production. Total meat trade across the globe increased from 10.6 MMT in 1993 to 26.1 MMT in 2013. The expansion of free trade, especially in agriculture, facilitated the global growth of the meat industry.

Third, increased corporate concentration, in large part facilitated by the WTO and other free trade policies, played a role in the expansion of the global meat industry. Corporate concentration—involving companies growing in size and gaining larger shares of markets, generally by acquiring or merging with other companies—occurred in a couple of ways in the global meat industry. First, the seed industry experienced significant corporate concentration as GE seed production increased. Howard (2009, 2016) demonstrates the tremendous concentration—in the form of mergers and acquisitions—that occurred in the global seed market from 1996 to 2008, with the commercial release of GE seeds. A handful of companies came to dominate this market during this period, particularly DuPont, Monsanto, and Syngenta. These three companies accounted for more than half of the global seed market in 2009. This consolidation centered on the development of GE seeds, as this new technology facilitated increased market concentration. The global meat industry also became more concentrated in terms of meat processors (Heffernan 1998; see also chapter 2). Heffernan (1998, 50) shows high levels of corporate concentration in meat processing in the United States: four firms controlled 55 percent of broiler chicken production, 87 percent of beef slaughter, and 60 percent of pig slaughter. In 2013, the top four pork processors controlled 63 percent of the market (Howard 2016, 83). As Howard (2009, 1270) points out, the reason behind this consolidation is fairly straightforward: "When concentration reaches a certain threshold, the largest firms are able to ensure stable profits by ceasing to compete on the basis of price." In the case of both GE seeds for feed grains and meat processing, the increased market concentration was accompanied by an expansion in the global reach of the corporations. Thus, GE seeds for feed grains came to be heavily adopted in several countries, especially Argentina, Brazil, Canada, and the United States. Indicative of the growing global reach, the import of GE feed grains occurs more widely today than twenty years ago and has come to include countries in Africa, Asia, Europe, the Middle East, and North and South America.[7] Consequently, as these markets expanded globally, a handful of companies grew in size, market share, and global reach.

While other factors have also contributed to the expansion of the meat industry, these three factors—increased feed grains production, a shift toward more liberal trade policies, and greater corporate concentration in the meat industry—have been particularly important for the growing global reach of the meat industry. Increased corn and soybean production have allowed for the growth and global spread of industrial livestock production that relies on intensive feed methods. This increased feed grain production, of course, has relied upon greater corporate concentration in the global seed industry with the spread of GE feed grains. Since resistance to GE foods existed in many parts of the world, most notably Europe, markets had to be opened, and this depended in part on corporate power and in part on a broader global shift toward free trade. Finally, this global shift toward free trade was pushed by corporations—including the feed grain industry in the United States—and countries that benefit from free trade (e.g., see Clapp 2003; Winders 2009, 2017). Thus, the three interconnected factors formed a foundation on which the meat industry could expand globally.

Consequences of the Global Meat Industry

The sharp increases in the global production and consumption of meat are not without profound consequences. As global corporate giants in meat processing have grown in size and reach, industrial animal farms have spread across the globe. Millions of acres of land have been converted to animal farming in the form of pastures, crops for feed, or as a reservoir for overwhelming volumes of animal waste; workers, animals, and the environment have all been negatively impacted. The expanding meat industry also has implications for issues such as world hunger. The reliance on feed grains—which diverts land and resources that might otherwise go toward food grains and legumes for direct human consumption—is one notable way that the global meat industry can contribute to food insecurity and world hunger, even as it contributes to greater food production. Of course, the food produced by the global meat industry is of higher economic value and therefore out of the reach of many people in poorer, food-insecure countries.

This growth of the global meat industry therefore generates three sets of concerns that undergird much of this volume: corporate concentration, the tension between producing cheap meat and reducing food security, and social and environmental injustices. First, there are political economic concerns regarding the size and power of corporations in the global meat industry. Global meat companies—such as Tyson, which is U.S.-based, and JBS, which is based in Brazil—exert significant influence over farmers, national governments, and many other groups. Unequal power between industrialized countries and poorer countries is also an economic concern, as some poorer countries produce and export substantial amounts of meat. Poorer countries producing large quantities of meat for industrialized countries can raise issues of especially weak environmental and labor regulations, as well as the issue of food insecurity by relying disproportionately on imports to meet domestic demand. Economic power shapes the entire meat supply chain, with animals reared in large concentrated animal feeding operations (CAFOs), while low-paid workers repetitively disassemble thousands of animals every day in massive industrial slaughterhouses dotting the global landscape from Australia to Brazil, from the United States to China.

Second, as the availability of meat from industrial production expands, smallholders often find themselves unable to compete with industrial production for market share and even for land. The loss of smallholder production can further contribute to a decline in both food security and food sovereignty for communities around the world. This competition with global industrial production also contributes to the disappearance of more traditional animals and foods, with consumers eating more uniform types of meat and grains, thereby contributing to less genetic diversity in the agrifood landscape.

Finally, there are important concerns about the environment, sustainability, and social justice. The growth of global meat production and consumption—the economic expansion of the global meat industry—has brought grave concerns about environmental consequences. Deforestation, methane from cattle contributing to climate change, pollution from CAFOs, and overfishing of our oceans are just a few of the many environmental impacts created by the expansion of industrial meat production globally. These concerns have been voiced by a myriad of organizations, including the Union of Concerned Scientists and the Food and Agriculture Organization of the United Nations (FAO) among many others. Also noteworthy is that the consequences of environmental degradation are rarely shared equally. Polluted water and air and more erratic weather conditions will disproportionally impact poorer nations and people, including ironically, smallholders who have been further marginalized by the global meat industry. Moreover, it is often poorer individuals who become the bulk of the labor force for CAFO operations and industrial slaughterhouses with the latter occupation having notoriously high injury and turnover rates.

The Outline of the Book

The chapters in this book bring empirical data to bear on a political-economic analysis and discussion of the global meat industry and its social and environmental consequences. The overall discussion rests on the understanding that there are many small-scale meat producers across the globe, but most of the meat consumed globally is from large-scale industrial meat production. This diversity in global meat production has implications for environmental and sustainability issues: small-scale meat production is a system in which animals are very important for providing not only draft power for plowing, but also fertilizer for gardens and field crops. By contrast, the global meat industry—centered on very large meat processing companies and concentrated industrial production—has the largest and most damaging environmental impacts seen within meat production.

Part I of the book explores global forces shaping the expanding meat industry, paying particular attention to the role of corporations and national governments. In chapter 2, Philip H. Howard examines the process of corporate concentration in the global meat industry by demonstrating how government subsidies have aided three of the biggest global meat processing firms—JBS, WH Group, and Tyson. These three corporations have rapidly increased in size and power in recent decades. Howard demonstrates how two kinds of government subsidies have aided these transnational corporations (TNCs) in gaining greater economic dominance: subsidies that reduce the cost of animal feed, and subsidies that reduce the cost of acquiring competitors. Tyson, which is headquartered in the United States, has benefited particularly from policies that subsidize corn and soybean production and thereby substantially reduce animal feed costs. WH Group, headquartered in China, has grown rapidly with the support of the national government, which has provided billions of dollars for buyouts. One of these buyouts was for the acquisition of Smithfield in 2013, which made WH Group one of the largest meat processing corporations in the world. Finally, JBS is headquartered in Brazil and has benefited from both feed and financial subsidies that have included access to billions of dollars from the Brazilian government to finance acquisitions. These cases demonstrate that government subsidies, both direct and indirect, have played a central role in the expansion of the global meat industry and the increased concentration within it, as well as the immense acquisition of wealth by the top executives of all of these firms.

In chapter 3, Conner Bailey and Nhuong Tran provide an overview of the global seafood sector—including traditional wild-caught fish (i.e., "capture") and farmed fish (i.e., "culture")—describing trends in production over the past several decades. They examine issues of environmental risk and social impacts in this sector of the global meat industry, demonstrating how such issues can pose a challenge to corporate concentration and dominance in the seafood industry. Bailey and Tran demonstrate that, despite such obstacles, global trade in fisheries products and the importance of feed in intensively managed aquaculture offer opportunities for corporations to increase their economic power within the seafood industry. They also demonstrate how control over breeding stock in seafood, including through genetic engineering in fish and shrimp, allow for the assertion of greater corporate control. Bailey and Tran therefore provide an important and informative examination of a sector of the meat industry that is often overlooked.

Part II of the book turns to particular places to show how the global forces propelling the meat industry are experienced on a more local level. In chapter 4, Mindi Schneider examines the dramatic rise in pork production and consumption in China's reform era (i.e., since 1978), and she demonstrates how this trend was intimately connected to the simultaneous expansion of the global meat industry. Schneider begins by discussing cultural history and meanings related to pigs, pork production, and peasants. Turning to the transformation of China's pork industry, Schneider argues that national policies and corporate strategies played key roles in developing the production system found in China today. In particular, the expansion of China's pork industry rested on national policies that liberalized agricultural markets, and facilitated industrialization and capitalization in agriculture beginning in the late 1970s. Schneider also demonstrates how such national policies connected to global forces, including the expansion of soybean production and trade, in ways that facilitated the expansion of the pork industry in China. In 2017, China had half the world's pigs, with about 690 million, and it produced and consumed half of the world's pork, at about 53 MMT. Schneider then shows how the country's current pork industry is both highly productive and environmentally destructive. Given its centrality to the global meat industry, understanding the growth of China's pork industry is especially important.

In chapter 5, Thomas K. Rudel explores sustainable cattle ranching practices among small-scale farmers in the Ecuadorian Amazon in South America within the context of the expanding global meat industry. Rudel explains why smallholders with livestock tend to engage in sustainable agricultural practices, even in contexts marked by globalization that works to undermine sustainable practices. This chapter provides a historical analysis of changing land management practices among two groups of small-scale cattle ranchers, one *mestizo* and the other Amerindian, in the Ecuadorian Amazon between 1986 and 2011. Drawing on surveys conducted over this twenty-five-year period, Rudel demonstrates that the intersection of globalization with racial and ethnic politics disadvantaged some landowners (Amerindians) and encouraged them to rent their land to mestizo cattle ranchers who degraded the pastures. At the same time, *mestizo* cattle ranchers in adjacent communities allowed silvopastoral landscapes—considered more environmentally beneficial—to emerge on their own lands in part because the spontaneous emergence of these silvopastures required minimal inputs of labor. By contrast, the rented lands of Amerindians have not developed into silvopastures landscapes. Rudel argues that these diverse land management strategies underscore the continuing importance of owner-occupancy in fostering environmental stewardship even in contexts, like Latin America, marked by globalizing markets.

In chapter 6, Carrie Freshour discusses the experiences of workers in the global meat industry, focusing on poultry plant workers in the southeastern United States. Based on ethnographic fieldwork in a slaughter plant, Freshour explores poultry work by focusing on conflicts between workers and corporations. She shows how corporate strategies have led to shifts in the composition of the workforce, in an attempt to undermine workers' power in demanding higher wages or better working conditions. Most recently, poultry processing plants have shifted from hiring primarily undocumented Latina immigrants to employing African American women. Freshour explains how this recent shift has unfolded, but she also demonstrates that historical continuity exists in the industry's reliance on marginalized workers, even though these groups have changed over time. Such shifts in the composition of the workforce in meat processing plants represent an ongoing conflict between workers and corporations, as corporations seek a cheaper, more docile, and more vulnerable labor force.

The third and final part of this book discusses some of the important environmental, social, and ethical consequences of the global meat industry, as well as some possible solutions. In chapter 7, Riva C. H. Denny examines the global meat industry's contribution to pollution and climate change. With its global expansion over the past several decades, the meat industry has become one of the most significant contributors to climate change because of the high levels of greenhouse gas (GHG) emissions produced by the tens of billions of animals raised for human consumption. Denny investigates both total GHG emissions and emission intensity (EI), which is the rate of GHG emissions per weight of meat produced. Using statistical methods, she examines GHG emissions from the meat industry drawing on FAO data for more than 190 countries from 1961 to 2014. Denny offers, then, a truly global analysis of the meat industry during this important period of immense expansion. Focusing on cattle, pigs, and chickens, Denny's analysis considers how different production techniques across different countries and regions contribute to global climate change.

In chapter 8, Robert M. Chiles and Celize Christy discuss the tensions between economic development and ethical considerations of animals. They focus their discussion on meat production in Rwanda, one of the world's poorest countries. In doing so, Chiles and Christy critically examine the assumption that intensifying livestock production is the best and most viable pathway to economic development and food security in countries of the global south, such as Rwanda. This view is often put forward by international organizations with livestock development programs, aiming to help countries in the global south, and especially their farmers. While such programs can help with food security, poverty alleviation, and similar issues, Chiles and Christy point out that the programs usually fail to give attention to animal welfare issues and humans' obligations toward animals. Similarly, the Rwandan government has created policies to encourage intensive meat production, particularly poultry. Chiles and Christy consider how such development policies affect animal welfare, and they discuss alternative pathways to achieving goals such as food security and poverty alleviation.

In chapter 9, Elizabeth Ransom and Bill Winders reflect on the overarching themes highlighted by the chapters: the influence of governments and corporations in facilitating and shaping the global expansion of the meat industry, the contributions of the expanding meat industry to food insecurity in many places around the world, and the resulting environmental and social consequences of this industry's transformation. Ransom and Winders explore some of the ways that individuals have responded to the global meat industry and its negative consequences. There are many individual consumer responses—such as reducing or eliminating meat consumption, or choosing local or organic meat. While such individual-level responses are important, they ultimately have difficulty adding up to viable and adequate solutions, given the larger forces behind the global meat industry. To that end, they note how organizations, governments, and even corporations can help to ameliorate some of the negative consequences of the global meat industry and begin to take steps toward alternative and more sustainable approaches.

The expansion of the global meat industry has important implications for our lives, including for the environment, people's access to food and other resources, economic control and power, the lives of animals, and even for our individual health. Understanding the expansion of the global meat industry over the past several decades is of paramount importance. And this expansion—in terms of the amount of production, the geographic reach, and the increase in trade—rests on the changes in the market structure. More specifically, the increased market concentration in the meat industry and the increased size of meat corporations over the past few decades have provided the foundation for the global expansion of the meat industry. So, we turn first to the role of governments and corporations in supporting this transformation.

Notes

1. Data on the value of global meat production comes from FAO 2019.

2. Denmark's declining per capita meat consumption was tied to that country's drive to decrease the use of antibiotics in pig farming in the 1990s. In just a matter of a couple of years, per capita pork consumption plummeted, from 68.5 kg in 1998 to 23.1 kg in 1999.

3. Beef in India comes primarily from buffalo rather than from cows, and its primary export markets are regional, including Vietnam, Thailand, Malaysia, Saudi Arabia, and the Philippines. See www.beefcentral.com/trade/export/where-does-indias-buffalo-meat-exports-go/.

4. Calculated from FAO 2019. "Russia" refers to the USSR from 1961 to 1991, and then the Russian Federation from 1992 to 2014.

5. Some of the corn that is produced is, of course, consumed directly as food, and some of it is also used for biofuel production, especially since 2008. Nevertheless, it is important to understand that the doubling of production of corn production

Introduction to the Global Meat Industry 23

since 1990 and the three-fold increase of soybean production was driven by expanding demand for livestock feed.

6. For example, chicken meat imports in Russia increased substantially after the fall of communism in 1991, from 0.2 MMT to 1.1 MMT in just a few short years, from 1993 to 1996.

7. For example, the EU had banned both the production and import of genetically engineered seeds until 2004, when it allowed the import of GE corn. Then, in 2008, it began to allow the import of GE soybeans (Winders 2017, 123–125). Almost all of the feed grains imported by the EU today are genetically engineered.

1 Global Forces

The global circulation of animals—either live, dead, or in parts—reveals important tendencies in the capitalist world economy, especially regarding corporate power and economic control. While expanding their geographic reach, transnational corporations (TNCs) in the meat industry have increased their profits and economic control by gaining greater market shares, acquiring competing corporations, and increasing their influence over other economic actors (e.g., farmers). Furthermore, governments have supported both the global expansion and the greater economic power of TNCs in the meat industry.

Although the beginning of industrial meat production is often traced to the founding of the Chicago Stockyards in 1893, the global meat industry exists due to the spread in scale and scope of industrial production techniques. Confined animal feeding operations (CAFOs) are a major component of industrial production, and these depend on large quantities of affordable livestock feed. Corn and soybeans, and to a lesser degree canola, are central ingredients in industrial livestock feed and are thereby the foundation of the global meat industry. The feed industry exhibits some of the same global dynamics found in the global meat industry, particularly regarding corporate power and economic control. First, the global production of feed grains—particularly corn and soybeans—has increased dramatically since 1960, and especially after 1990. This global increase in production occurred as both chemical and grain corporations expanded their global reach. Governments supported this global expansion in important ways, most notably by subsidizing production. Second, chemical corporations increased their economic power through the development of genetically engineered (GE) seeds, which contributed to more concentrated markets and more corporate control over inputs that farmers rely upon. We look briefly at the trends in the global feed industry to reveal some insights about tendencies in the capitalist world economy and the global meat industry, as well as to highlight the resistance that has emerged to the global expansion of the feed industry.

First, a tremendous expansion occurred in the production of feed during the past few decades. From 1990 to 2017, global corn and soybean production more than doubled—from 481 MMT to 1,033 MMT for corn, and 104 MMT to 336 MMT for soybeans (e.g., see Winders 2017, 3, table 1.1).[1] In 2016, the United States produced 384 MMT of corn, with China second at 219 MMT, followed by Brazil at 98 MMT, the EU at 62 MMT, and Argentina at 41 MMT of corn. Similarly, the United States produced 119 MMT of soybeans, followed by Brazil with 114 MMT, and Argentina and China produced 58 MMT and 12.9 MMT, respectively. The United States and Brazil, then, have been at the forefront of the global expansion of soybean production; and the United States, China, and Brazil have driven the expansion in global corn production.

Part of the increase in the global production of corn and soybeans was tied to an expansion of land used in producing these commodities. From 1990 to 2016, worldwide corn production expanded from 129 million hectares to 189 million hectares, and soybean production increased from 54 million hectares to 119 million hectares.[2] China led the expansion of land used for corn production, while South America led in the expansion of land used for soybean production. Corn area harvested in China increased from 21.4 million hectares in 1990 to 36.7 million hectares in 2016. In the same period, corn area harvested increased in Argentina rose from 1.9 million hectares to 4.9 million hectares, and Brazil from 13.5 million hectares to 17.6 million hectares harvested. South American soybean production had a larger expansion in land harvested in this period: Argentina went from 4.7 to 17.4 million hectares harvested, and Brazil increased from 9.7 to 33.9 million hectares. By contrast, the number of soybean hectares harvested in China decreased during this period, from 7.5 to 7.2 million hectares. The United States saw an expansion in corn and soybean production, as well, from 27 million hectares of corn to 35.1 million hectares, and from 22.8 million hectares of soybeans to 33.4 million hectares. More than half of the global expansion in soybean hectares, then, occurred in Argentina, Brazil, and the United States.

This expansion in the global production of corn and soybeans fueled an increase in international trade in these commodities. Argentina, Brazil, and the United States account for the majority of corn and soybean exports in the world. The United States is the world's leading exporter of corn, with Brazil second and Argentina third. And Brazil is the leading exporter of soybeans, with the United States second and Argentina third. The United States and Brazil, then, dominate the world trade in feed grains. Most of the feed grain from the western hemisphere is destined for Europe, Asia, and northern Africa. This trade pattern in feed grains, then, is intimately tied to the expanding meat industry in these regions, which have seen meat production increase significantly.

Governments and TNCs have been central to this increased global production and trade of feed. First, governments have liberalized national agricultural policies, which had attempted to limit agricultural production through supply management programs (Winders 2009, 2017). This shift in national polices has allowed many farmers across the globe to begin producing feed grains where they had previously faced production controls. At the same time, governments have worked to liberalize trade, as free trade agreements have proliferated (Winders et al. 2016). Free trade agreements, such as NAFTA, and the creation of the WTO in 1995 have removed trade barriers that had previously restricted trade in corn and soybeans. Therefore, liberalization in national agricultural policies and trade policies has contributed to both producing and trading more feed grains.

Second, TNCs have played a role in the increase in the global production and trade of feed grains, in part by gaining greater control over markets. Just a handful of grain corporations have come to dominate trade in the global feed industry: Archer Daniels Midland (ADM), Bunge, Cargill, and Louis Dreyfus. These corporations control more than "70 percent of the global grain market, though they face growing competition from new companies in Asia, including Noble Group, Olam, and Wilmar, which are three Singapore-listed agribusinesses; Cofco in China; and Glencore Xtrata in Switzerland" (Winders 2017, 7). Similarly, a handful of companies control markets for key inputs in feed grains, such as seeds. In 2015, just a few chemical corporations controlled about 70 percent of the global seed market: BASF, Bayer, Dow, Dupont, Monsanto, and Syngenta (Turzi 2017, 26). Corporate concentration has been rapid over the past few decades in the seed industry (Howard 2009), as Monsanto and other leading seed corporations have acquired or merged with numerous smaller seed and chemical companies. Most recently, the seed market has become even more concentrated as Dow and DuPont merged in 2017, and Bayer acquired Monsanto in 2018. Such increased market concentration has had two effects. First, the larger companies have more resources with which to expand their global reach. Second, the mergers and acquisitions behind such market concentration often involve corporations from multiple nations—such as Bayer (Germany) and Monsanto (United States)—which again increases the global reach of the corporation. In both of these ways, then, greater market concentration contributed to the emergence of a truly global feed industry.

In addition to this geographic expansion of corn and soybeans, the development of GE corn and soybeans has also fueled the global expansion of the meat industry. Globally, GE production has expanded from 1.7 million hectares in 1996 to 181 million hectares in 2014 (James 2014). Importantly, though, just four crops account for more than 90 percent of GE production worldwide: canola, corn, cotton, and soybeans. First commercially available in 1996, three of these GE crops—canola, corn, and soybeans—are central ingredients for feed grains. The top three countries in terms of GE planting—Argentina, Brazil, and the United States—are also the leading exporters in corn and soybeans. More than 90 percent of corn and soybean production in these countries is GE. Therefore, the expansion of global corn and soybean production was driven by the adoption of GE seeds. And this expansion of feed production from GE seeds has, in turn, extended corporate control over farmers by adding legal power because the seeds are patented. For example, chemical corporations have required farmers to sign technology agreements that limit farmers' ability to save seeds, include a technology fee, and set a premium price for the seeds (Eaton 2013; Kinchy 2012). All of this, of course, works in favor of the corporations making a big profit.

These changes in feed grain production—the geographic expansion and the rise in GE seeds use—have contributed to greater consolidation and concentration in farming that have imperiled the very existence of some farmers. Guptill and Welsh (2014) explain that Canada and the United States have seen a decline in mid-size farms and an increase in large farms. In Argentina and Brazil, the expansion of soybean production has led to land conflicts between peasants and large landowners, with some peasant leaders being murdered (Lapegna 2016, 37–39; Turzi 2017, 87). Even without such violence, the global expansion of corn and soybean production threatens many farmers by contributing to declining prices in the long term that lead farmers to produce more (e.g., by adopting GE seeds or expanding the amount of land farmed) or to leave farming altogether.

Such threats have provoked resistance in various ways, as many farmers in the global north and smallholders in the global south have challenged these changes in feed grain production. In the global north, farmers have organized against GE crops in various ways. In 2004 and 2005, for example, members of the French farmer organization Confederation Paysanne engaged in "crop-pulls" in which activists uprooted GE crops in several Monsanto open-air field trials. In Canada in 2002, organic canola farmers filed a class action lawsuit against Monsanto. In the global south, smallholders have often mobilized to protest these changes. In Argentina, for example, peasants "organized several roadblocks and filed suits against agribusinessmen, demanding reparations for damages to their farms" when glyphosate herbicide used with GE soybeans drifted to their fields and destroyed the peasants' cotton and vegetable crops (Lapegna 2016, 1). Despite such resistance from farmers and smallholders, the production of GE feed grains continues to spread across the globe, very much in tandem with the growth of the global meat industry.

The development of the global meat industry has depended on the expansion of global feed production, and vice versa. Turzi (2017, 19n10) points out clearly the important connection between production in each of these industries: "Small increases in per capita meat consumption—in the context of feedlot dominance—will lead to large increases in demand for feed proteins." And, both the global meat industry and feed production have rested on three key factors: (1) the liberalization of the world economy, making trade easier for businesses, including expanding trade and production networks; (2) the development of technology that supports market concentration and corporate economic power; and (3) support from national governments to encourage, facilitate, and even subsidize the global expansion.

The chapters in part I help to illustrate and analyze these global forces. First, in chapter 2, Howard discusses the role of TNCs in the global meat industry. Howard offers a case study of three corporations central to the global meat industry—JBS of Brazil, Tyson of the United States, and WH

Group of China. Through these case studies, Howard shows how states have supported corporations in their drive to expand globally through different kinds of subsidies. Next, in chapter 3, Bailey and Tran illustrate how environmental risks and risks in the production process inhibit increased market concentration and corporate power in the global seafood industry. Yet, they also note how three factors—international trade, feed production, and biotechnology (i.e., genetic engineering)—might allow for greater corporate concentration and power. These chapters, then, examine global forces that have shaped the expansion of the global meat industry over the past several decades.

Notes

1. Data on feed production comes from USDA FAS n.d.
2. Ibid. Data is from "area harvested."

2 Corporate Concentration in Global Meat Processing: The Role of Feed and Finance Subsidies

Philip H. Howard

Like many other industries, the meat processing industry has become much more global, and ownership has concentrated dramatically in recent decades. The three largest firms globally are JBS of Brazil, Tyson Foods Inc. of the United States, and WH Group Limited of China. When their shares are aggregated in the U.S. market they control 63 percent of pork packing, and just two (Tyson and JBS) control 46 percent of beef packing and 38 percent of poultry (Tyson Foods 2016). These levels are of concern because institutional economists describe a market where four firms control 40 percent or more of sales as an oligopoly, or a shared monopoly, due to conducive conditions for increasing prices (Howard 2016). In addition, the largest meat processors have concentrated markets not just (1) horizontally, through acquiring direct competitors in their initial processing sectors (e.g. poultry); they have also grown (2) concentrically, by branching into the processing of additional livestock species; and (3) vertically, by taking over upstream suppliers (e.g. animal genetics, feed mills, feedlots) and downstream packaged/branded food manufacturers. By 2018, among all global packaged food firms, the largest included JBS ranked at number 2, Tyson at number 3 and WH Group at number 13 (Kalkowski 2018). These meat processors each control dozens of brand names, giving retail consumers the illusion that ownership remains quite diverse. Even more hidden from public view, however, the meat processors are reshaping a system that was previously characterized by a long series of stages/markets between farmers and consumers, each composed of numerous competitive firms. In its place, they are moving toward an increasingly "seamless system," with just a few firms controlling every aspect of production (Heffernan, Hendrickson, and Gronski 1999).

Government subsidies, both direct and indirect, have been crucial in supporting these trends. These are numerous, and include direct ownership stakes, payments or tax breaks for production, low-interest funds to finance acquisitions, policies that shift the burden of environmental and community impacts of their operations to taxpayers, and regulatory barriers that disadvantage competing firms. This chapter focuses on two key subsidies that reduce the costs for (1) obtaining the key input of animal feed, and (2) acquiring competitors. Such advantages have greatly assisted the leading firms in overcoming previous limits to global dominance, both biophysical and social.

This chapter also explores the tensions between "legitimation and accumulation" (O'Connor 1973), as government efforts to assist the most dominant firms also threaten to undermine public support—not only for subsidies, but also for the authority of governments themselves. Political and economic changes in the meat processing sector, for example, have restructured societies in ways that have led to the loss of livelihoods of numerous smaller packagers, processors, farmers, and breeders. In addition, they have resulted in negative impacts for public health, animal welfare, and local and global ecosystems. Furthermore, they threaten the resilience of the food supply by locking in a highly centralized system that is increasingly vulnerable to potential disruptions, such as climate change and disease outbreaks. Both governments and dominant firms seek to justify these consequences through rhetorical strategies, and by efforts to obscure the full costs of their actions.

The next section reviews the literature on government subsidies and their role in facilitating global corporate concentration. This is followed by case studies of the three largest meat processors globally, with a focus on data from the most recent twenty-year period (1996–2016). The analysis compares the different strategies that national governments have used to fuel and rationalize the growth of these firms, as well as the role of these firms in garnering these supports. It suggests that Tyson has benefited greatly from subsidies for animal feed, WH Group from finance subsidies, and JBS from both types of subsidies, which has helped the latter to achieve its current ranking. The concluding section then highlights the impacts of these trends, and then explores the likelihood that they will continue.

Global Concentration and National Government Subsidies

Decision-makers in dominant firms are constantly seeking to increase the power of their organizations (Nitzan and Bichler 2009). If they do not, they are likely to lose investors or become vulnerable to takeover by other firms, or both. Governments frequently assist these efforts (Baran and Sweezy 1966), as their policy changes are overwhelmingly shaped by elites (Bartels 2010; Gilens 2014; Schlozman, Verba, and Brady 2012). O'Connor (1973) described the challenges that governments face, however, as they negotiate tensions between helping dominant firms to accumulate more power, and a loss of legitimacy as this process negatively impacts the rest of society. Corporations increasingly recognize the importance of maintaining broad support for the political economic system, and are devoting more resources to public relations, funding for think tanks and endowed faculty positions, and other methods of reducing the potential for resistance (Boyd 2000).

Government subsidies often give dominant firms advantages over competitors, and even small gains can become magnified as the "rich get richer" (Barabási and Bonabeau 2003; Easley and Kleinberg 2010), thus reinforcing trends toward concentration. As leading firms have encountered limits to increasing power in their nations of origin, they have expanded globally in search of cheaper inputs and additional markets (Constance and Heffernan 1991). This greater scope provides even more advantages over competitors, as they are able to pit nation-states against each other, scouring the globe for the most favorable government supports and the weakest regulatory oversight (Bonanno and Constance 2010). A global scope also makes tensions between legitimation and accumulation more visible, as the governments that fostered the rise of these firms find it even more difficult to rationalize the benefits for their own bureaucracies, let alone the majority of their citizens.

Government actions to support accumulation by dominant firms have been essential in overcoming both biophysical and social barriers to their growth. Biophysical limits have been overcome through government actions that subsidized and artificially cheapened the costs for feed and for long-distance transportation, as well as increased the ability to produce livestock in more confined spaces (government-funded research for faster growth, reduced feed consumption, disease control, disposal of more concentrated wastes, etc.). Social limits have been overcome through actions that, for example, helped to give dominant firms advantages over competing firms (easier access to finance, higher direct subsidies, lower levels of regulatory oversight, increased barriers to entry for smaller firms), reduced labor costs (anti-union "right-to-work" laws, more government assistance to supplement below-subsistence wages) (see chapter 6 for discussion of low-wage work), and reshaped dietary patterns (increased consumption of meat and more highly processed meats).

For dominant firms that seek to grow faster than competitors, Nitzan and Bichler (2009) identified two key pathways: pursuing *depth*, which involves raising prices or reducing costs, and *breadth*, which involves internal or external growth. Subsides that enable dominant firms to employ these strategies more effectively or with less risk than their closest competitors may provide critical advantages.

Some of the most important subsidies in meat processing are those that reduce animal feed costs and finance acquisitions. Feed, for example, is typically the largest cost embodied in the retail price of meat, accounting for 60 to 70 percent of expenses in confined production systems (van Huis et al. 2013, 171). In addition, internal growth is difficult for processors focused on markets in the global north, where per capita meat consumption has leveled off. The ability to achieve external growth through acquisitions (figure 2.1), especially if financing can be arranged at low interest rates, can significantly influence which firms (and their executives) rise or fall.

Not surprisingly, overcoming previous biophysical and social limits to the growth of dominant meat processors has resulted in substantial negative impacts, not only the intended effects of increasing social inequalities, but also the collateral damage of additional human and environmental costs (Cochrane 2010). These "externalities," which are not typically calculated in economic analyses, include the loss of viable rural communities, the public health impacts of increased consumption of industrially produced and highly processed meats, resource depletion, pollution, and the increased suffering of animals. Meat processors and government actions increasingly are delinking livestock production from its previous ties to nearby land resources (e.g., sources of feed), enabling the concentration of much higher numbers of livestock in increasingly small areas (Naylor et al. 2005).

Corporate Concentration in Global Meat Processing

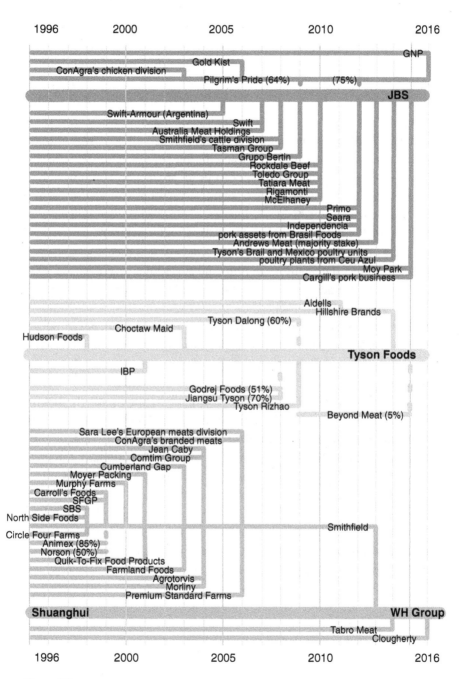

Figure 2.1
Leading global meat processing firms: timeline of ownership changes, 1996–2016

One example of the increasing socioeconomic and environmental interactions between distant places is the exponential rise in shipments of soybeans from Brazil and the United States to China for livestock consumption (Liu et al. 2013). This is not only a process of concentrating ownership and production, but a process of separating people from their means of production, and livestock from feed and nutrient cycles (Schneider 2017b). These actions to reshape society and ecosystems increase the "ecological hoofprint" of meat by constantly increasing the amounts of external inputs like energy, water, nutrients and chemicals in its production and distribution (Weis 2013a,b). In addition, these trends, which have also led to greater genetic uniformity of livestock, increase the risk of disease outbreaks in both animals and humans (Manning, Baines, and Chadd 2007).

These impacts are rationalized or obscured through rhetorical strategies, including direct statements from firms and government agencies, as well as proxies, such as trade associations, think tanks, fake grassroots ("astroturf") groups, and so on (Hamerschlag, Lappe, and Malkan 2015). Some common themes include appeals to the "efficiency" of an increasingly centralized system and the need to feed growing populations. The illusion of efficiency is reinforced by relatively cheap prices for consumers (Carolan 2014), but that can only be maintained if subsidies and damages are ignored (Weis 2010, 2013b). Claims of feeding hungry populations are similarly shaky—the increasing throughput of industrialized systems does not address distributional issues, which are exacerbated by these trends (e.g., undermining informal food systems and displacing smaller, less resource-intensive producers) (Schneider 2014). What Freudenberg (2005) calls "diversionary reframing" (or more plainly, changing the subject) is also a common strategy, such as deflecting blame for these problems toward other actors. Industries and governments have attributed recent outbreaks of avian influenza to wild birds, for example, rather than large-scale confinement operations (Wallace 2017), although smaller, outdoor operations have been less affected than industrial operations (Philpott 2015).

Feed Subsidies: Tyson

Tyson Foods is the second largest meat producer in world and the largest in the United States. The publicly traded corporation is ranked first in processing of poultry and beef, with U.S. market shares of 21 percent and

24 percent, respectively. It is also ranked second in pork, with an 18 percent market share (Tyson Foods 2016). Tyson initially was founded as a poultry firm in the 1930s by John W. Tyson, shortly before this industry began rapidly consolidating and vertically integrating.

Tyson receives a diverse array of subsidies, but among the most important are those that reduce the costs of corn and soybeans that are fed to livestock. In the United States, Department of Agriculture programs providing direct payments to farmers of these crops enabled Tyson to save an estimated $288 million per year just for its chicken division, according to an analysis of data from 1997 to 2005 (Starmer, Witteman, and Wise 2006). These calculations do not include the opportunity costs of growing government-subsidized animal feeds, rather than crops for direct human consumption, which would require significantly fewer resources per calorie (Winders 2017). Nor do they include the ecological impacts of growing crops that require substantial fertilizer and pesticide inputs, such as the zone of hypoxia (low oxygen) in the Gulf of Mexico that results from the runoff of these inputs and suppresses aquatic life.

The industrial model of animal agriculture was first developed for chickens, which are smaller and reach maturity faster than other key livestock species. This model is now being effectively applied to pork. Barriers to its application in beef systems are stronger but are also slowly being dismantled (e.g., via growth promoters such as ractopamine). Significant differences in feed conversion efficiency persist, however, averaging 1.7 pounds of feed to produce a pound of body mass in chicken, compared to 2.9 pounds of feed for pork, and 6.8 pounds of feed for Hereford beef (Bourne 2014).

This currently gives advantages to firms specializing in species with the most efficient conversion rates, particularly in geographic regions where feed costs are lower than in other parts of the world. Examples include poultry and pork firms in the United States, Brazil, and Argentina that have cheaper access to soybeans than firms in China and the EU. For firms specializing in less efficient species, such as beef, geography can also provide advantages, such as those in Brazil, Argentina, and Australia that make use of less expensive pasture (all year-round) for the majority of their cattle feed.

Tyson is not the only firm in the United States to have the advantage of subsidized animal feed for pursuing growth through *depth*, but its executives were willing to take more risks than competitors to achieve *breadth*. The founder's son, Don Tyson, convinced his father to continually reinvest revenues in expansion when most of their competitors did not. In what is considered a low-margin industry, the firm was able to better withstand market cycles, as well as to acquire competing firms when poultry prices dropped (Leonard 2014). For example, in 1998 Tyson acquired Hudson, another large meat processing firm, at a relative bargain price of $682 million (aided by an outbreak of E. coli and a subsequent USDA-ordered recall that some analysts viewed as excessive). Even so, founder James Hudson's family received $515 million to exit the business (Warner 1997).

Tyson's corporate motto is "segment, concentrate, dominate" (Bonanno and Constance 2010, 132), and Tyson considered it important enough to file a trademark application on these three words in 1998. Its executives follow this motto by identifying narrow markets on which to focus their resources, increasing the firm's share of these markets to achieve the top ranking, and then selecting other segments to dominate. Figure 2.2 shows the location of Tyson acquisitions over a recent twenty-year period. In 2001, for instance, the firm acquired IBP for $4.7 billion, after winning a bidding war with Smithfield. This move helped Tyson to achieve a dominant market share in beef processing, and a temporary position as the world's largest meat producer and processor (Barboza and Sorkin 2001). Tyson's geographic expansion has focused primarily on Asia through several joint ventures with firms in China and India, and some of these have been converted into full ownership.

Tyson's upstream subsidiary Cobb-Vantress supplies breeding stock for its operations, and it has acquired or partnered with a number of other poultry genetics firms in recent years, as shown in table 2.1. An estimated 95 percent of the world's commercial breeding stock for chickens produced for meat is now controlled by Tyson and just two other firms: EW Group and Groupe Grimaud (IPES-Food 2017). This has resulted in a high degree of genetic uniformity, combined with highly concentrated production, which enables diseases to spread more easily (see chapter 3). Tyson and other firms in the United States and Asia have experienced epidemics of avian influenza that have impacted poultry breeding and production (although some costs have been offset by hundreds of millions of dollars in subsidies for disease control, as well as government payments to growers who experienced losses) (Greene 2015).

Table 2.1
Ownership change for Tyson's Cobb-Vantress subsidiary

Date	Ownership change	Location
2007	Acquisition of Hybro (from Hendrix)	Netherlands, EU
2008	Joint venture for R&D with Hendrix	Netherlands, EU
2008	Acquisition of Avian Farms	Maine, United States
2008	Partnership with Sasso	France, EU
2013	Majority stake in Hubei Tong Xing	Hubei, China
2014	Acquisition of Heritage Breeders	Missouri, United States

Don Tyson, who died in 2001 at age eighty, was by that time a billionaire and one of the four hundred richest people in the United States. Yet most of the people employed directly by Tyson are no longer unionized and receive extremely low wages (see chapter 6). A sign in a Tyson processing plant in Arkansas states in both English and Spanish: "Democracies depend upon the political participation of its citizens, but not in the workplace" (Striffler 2002, 306). Tyson also exploits the labor of farmers who raise the company-owned poultry through its use of contracts. The firm uses a tournament system, which it pioneered, although other processors have adopted the system (and it has extended to the pork industry). In this system contract farmers are ranked against each other using nontransparent information, and their compensation is adjusted to reward high performers and punish low performers. Many variables such as the quality of feed and livestock are out of the farmers' control (Domina and Taylor 2009). Although Tyson's power has increased substantially over time, the benefits have not trickled down to the rural communities in which it operates. The journalist Christopher Leonard (2014), for example, found that per capita income in the majority of the seventy-nine counties where Tyson has facilities did not grow as fast as the state average.

Tyson and other U.S. meat processors have lobbied extensively to protect crop subsidies, and they have also lobbied against subsidies for ethanol that have diverted feed crops for use as fuel and slightly increased the price of feed in recent years (Kabel 2006). In response to pressure to end direct payments, these programs have shifted to subsidies for crop insurance, including insurance against market price declines, resulting in very similar

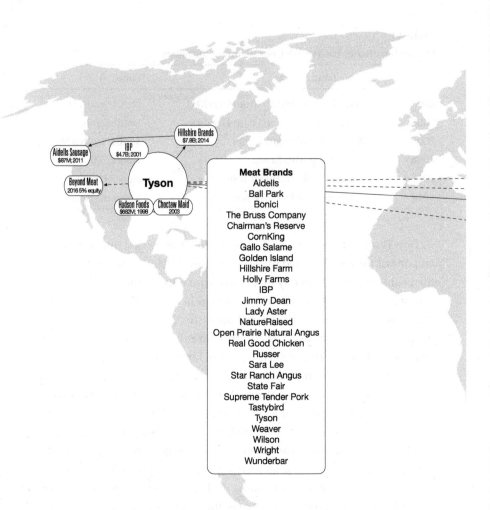

Figure 2.2
Tyson: ownership changes, 1996–2016

impacts for keeping feed prices low. Tyson also funds a number of organizations that attempt to put a positive spin on the corporation's increasing dominance, including the Center for Consumer Freedom, Center for Food Integrity, and the AgChat Foundation (Hamerschlag, Lappe, and Malkan 2015). The corporation's website touts its commitment to animal welfare, sustainability and charitable giving. Tyson also promotes its image with advertising, which is a tax-deductible expense.

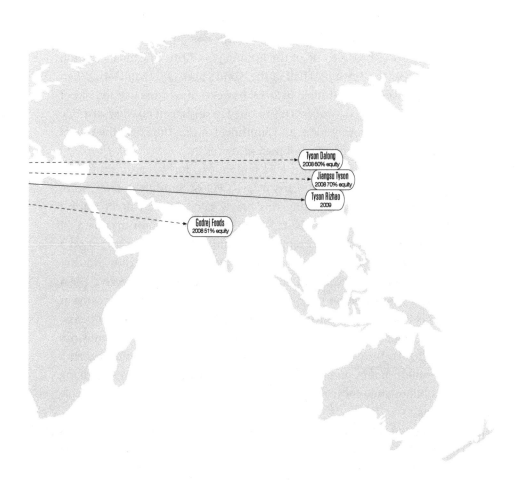

Figure 2.2 (continued)

Finance Subsidies: WH Group

In comparison to Tyson, WH Group has benefited far more from government finance subsidies than subsidies for animal feed. WH Group is the world's largest pork processor, and it is also dominant in the United States with a 25 percent market share (Tyson Foods 2016). The firm was started as the government-owned Luohe Slaughterhouse in the province of Henan, China, in the late 1950s, but it went bankrupt in 1984. It was then reorganized with the appointment of Wan Long as general manager of the factory (Tao and Xie 2015). The firm was later renamed Shuanghui and grew to become one of the leading pork firms in China, a nation that produces and consumes half of the world's pork (see chapter 4). When Shuanghui acquired Smithfield in 2013, however, it was just half the size of the latter, as measured by sales. This resulted in the largest takeover of a U.S. firm by a Chinese corporation, and Smithfield at that time held the title of the world's largest hog raiser and pork producer.

The Bank of China provided a $4 billion loan for this acquisition and approved it in just one day. Larry Pope, CEO of Smithfield, after being shown the documents detailing this move, said, "Wow. I don't think I could go out today and get the U.S. government to support making a $4 billion loan as a social responsibility for Smithfield to move forward on a foreign ... country's territory. No, I don't think that's doable in any industry that I can think of" (Woodruff 2014). Pope was expected to receive $46.4 million in compensation upon the sale of the corporation, which came as Smithfield was under criticism for its high level of executive compensation and poor performance relative to firms in closely related industries (Smith 2014). Wan Long remains head of WH Group and paid himself a $460 million bonus after acquiring Smithfield (Halverson 2015), making him a billionaire—he is now one of the four hundred wealthiest people in China.

Shuanghui subsequently reorganized under the name WH Group. It has continued to expand globally with government support and has since made more acquisitions in the United States, as well as Australia (figure 2.3). The firm is planning further acquisitions with a goal of becoming world's largest packaged meats firm, and it is expected to reduce the cost of financing for their Smithfield subsidiary to allow it to be the face of many of these transactions (Sito 2016). WH Group is rapidly expanding from its origins in pork processing to vertically integrated poultry production. One example is a joint venture with Nippon Ham Japan to supply technology for facilities in Henan province that will initially produce 50 million chickens per year (Pi 2014). The higher feed-conversion efficiency of chickens will allow the firm to produce more meat at a lower cost than its pig operations in China, although it will also require shifting strong cultural preferences for pork toward this alternative.

It is notable that Smithfield benefited from USDA feed production subsidies prior to its acquisition by WH Group—these were estimated to have saved $284 million per year for its hog division from 1997 to 2005 (Starmer and Wise 2007). Smithfield was also very aggressive in making acquisitions before its growth slowed and the firm became vulnerable to takeover. In addition to buying a number of its U.S. competitors, Smithfield took advantage of Poland and Romania's admittance into the European Union in the early 2000s to privatize previously government-owned firms there. It also established front companies to acquire more farms to circumvent limits to foreign ownership (Public Citizen 2004). In Romania its expansion coincided with the loss of 90 percent of hog farmers (400,000 farms) in just four years (Carvajal and Castle 2009). Smithfield also received $40 million in loans from the Word Bank to expand its Norson joint venture in Mexico, just a few months before being acquired by WH Group.

After WH Group acquired Smithfield, its new subsidiary immediately began exporting cheaper U.S.-raised pork to China. Imports in China have increased significantly since 2007 when it became a net pork importer, reaching a record high in early 2016 (Gale 2017). One limiting factor in China is the lack of sufficient arable land for feed crops to supply industrial pork farms, so China has increased its imports of corn and soybeans in an attempt to overcome these limits (Peine 2013; Schneider 2011). When including subsidies for other aspects of production (such as a strategic pork reserve, grants, subsidized loans, and tax breaks), the pork industry in China receives an estimated $22 billion, or $47 per pig (*Economist* 2014). Despite this, the USDA estimates that it is currently cheaper to produce hogs in the United States than China, with feed costs playing the largest role (Philpott 2013). WH Group, through its division Smithfield Grain, is also vertically integrating its supply chain for animal feed in the United States to reduce costs further. The firm is purchasing grain elevators in Ohio and securing contracts with grain farmers to supply the majority of its feed inputs, as well as ending previous contracts with many grain traders (Hirtzer 2016).

Also limiting the growth of pork production in China are the environmental impacts and lack of regulatory enforcement of existing operations. One example was 16,000 dead pigs that were found dumped into the Huangpu River in 2013, after a crackdown on sales of dead meat—the animals were affected by porcine circovirus (Davison 2013). Incidents

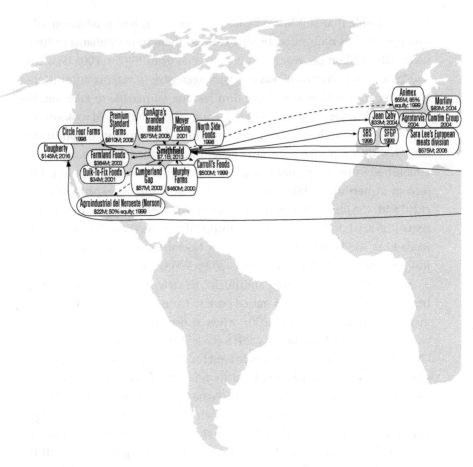

Figure 2.3
WH Group: ownership changes, 1996–2016

like these have resulted in increasing restrictions on production near rivers and more affluent areas and have contributed to insufficient domestic supplies to meet the growing demand for pork. A 2011 national government directive detailed a five-year plan to acquire foreign businesses to help address this issue, which justified providing Shuanghui/WH Group with the resources to acquire a larger U.S. firm (Halverson 2015).

The Chinese government has also rationalized the growth of WH Group and other large meat processors with rhetoric of efficiency and rural development (Schneider 2017a). Following the U.S. industrial model, government policies specifically encourage economic and geographic concentration by

Corporate Concentration in Global Meat Processing 45

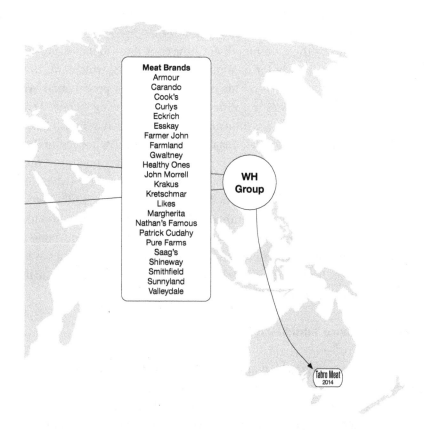

Figure 2.3 (continued)

designating "dragon head" enterprises—the term refers to dragon dancers in parades, and the importance of the person who wears the head of the creature for those who follow in line as its body. This leadership role has been assigned to select firms in a number of agribusiness industries (e.g., COFCO for grain trading, New Hope Group for animal feed) to facilitate increasing the scale of production, as well as more vertically integrated supply chains. Dragon head meat processing firms are technically obligated to contract with household-level producers to address rural development goals, but frequently inflate these numbers, and instead work with larger-scale producers (Schneider 2017a). These policies have been effective in dramatically transforming the pork sector in China, reducing smallholder producers from 74 percent of pig production in 2000 to less than 37 percent just a decade later (Schneider and Sharma 2014).

These actions by the government of China contribute to increasing pork consumption, particularly for its more affluent citizens, which aids in its legitimation (Schneider 2014). This is an example of a process that Weis (2013b) describes globally as "meatification," and is spurring increased consumption of meat in other less industrialized nations such as Brazil, Russia, and South Africa. The ecological, social, and human health impacts of these trends are downplayed by both WH Group and the government of China. Smithfield's slogan, for example, is "Good food. Responsibly" (which is a registered trademark). The corporation has an annual sustainability report that highlights its initiatives in animal welfare, the environment, and food safety. Its parent corporation, however, is not part of the Round Table on Responsible Soy (discussion follows), and the government of China has not encouraged any firms headquartered in the nation to participate in this organization (*Economist* 2014).

Feed *and* Finance Subsidies: JBS

JBS is the world's largest meat processor, a position that was greatly aided by both feed and finance subsidies. The firm is dominant globally in beef and poultry processing but is ranked second for both in the U.S. market, with market shares of 22 percent and 17 percent, respectively. It is also ranked third in the United States for pork processing, with a market share of 20 percent (Tyson Foods 2016). The firm was founded by José Batista Sobrinho in 1953, with a focus on slaughtering beef, and was renamed after his initials in 2005—it was initially called Friboi.

The dramatic growth of JBS in recent years coincided with receiving access to low-cost loans from the Brazilian government, in exchange for becoming a shareholder. This reflects Brazil's "national champions" development strategy, which resulted in government investments in some of its largest firms, and particularly in the meat sector, due to its world-leading position in exports of these products (Pigatto and Pigatto 2015)—other industries of focus included beer (Ambev/InBev), iron ore (Vale), telecommunications (Oi), and petroleum (Petrobras). Brazil's development bank, BNDES, acquired a stake in JBS in 2007, which currently accounts for 20.36 percent of shares, and another government-owned bank, Caixa, has a

4.99 percent stake. The total government investment was once as high as 31.41 percent (Degan and Wong 2012), but is being reduced to avoid being affected by the nation's credit rating. BNDES also made smaller investments in the Brazil-based meat processor Marfrig, which enabled it to acquire firms in the United States and UK (Pigatto and Pigatto 2015), and financed a merger of two leading firms to create Brasil Foods (BRF)—Marfrig and BRF are also ranked in the top ten among global meat processors (Sharma and Schlesinger 2017).

JBS has faced increasing criticism for receiving unfair advantages from the government of Brazil. In 2015, for example, it fell under investigation by the federal accounts court, questioning the "privileged treatment granted to the company," such as the speed with which loans by BNDES were approved for very complicated and risky acquisitions (Clarke 2015). In 2016, two of the founder's sons, Wesley Batista and Joesley Batista, temporarily stepped down from their roles as CEO and chairman of JBS, respectively, due to being detained (and later jailed) in an investigation of pension fund fraud, which targeted other firms held by the family (Magalhaes and Jelmayer 2017). Although state-run firms are not allowed to finance political campaigns, the government's minority stake allows JBS to spend more on candidates than any other firm in Brazil, including nearly a third of the members of the chamber of deputies.

Sergio Lazzarini, a professor of organization and strategy at a Brazilian university said, "The company has invested a lot in the management of the political interface," and "would JBS have had this success without the support of state capital? Probably not" (Schmidt 2014). Joesley Batista admitted as much in 2017, after receiving immunity from some criminal charges (resulting from an investigation of the alleged sale of tainted meat) in exchange for his testimony. Joesley disclosed payments of $220 million in bribes to thousands of politicians, and said that without these actions, the growth of JBS "wouldn't have worked. It wouldn't have been so fast" (Freitas, Freitas, and Wilson 2017). The future growth of JBS is now threatened, as the firm was assessed a fine of $3.1 billion for admitting to these bribes. It is in the process of selling off some assets to pay this penalty, such as feedlots in the United States and Canada. Nevertheless, five of six Batista's children hold investments in the company, and each of the five are now billionaires (Schmidt 2014).

Figure 2.4 shows major ownership changes involving JBS from 1996 to 2016, totaling more than $20 billion. This included acquiring the much larger U.S. firm Swift in 2007, which had become vulnerable to takeover due to declining profitability. Swift was itself the result of ConAgra acquiring Swift and Montfort, before eventually selling to the private equity firm Hicks, Muse & Co. in 2002. Plans to quickly sell the business for a large profit were thwarted by the outbreak of bovine spongiform encephalopathy

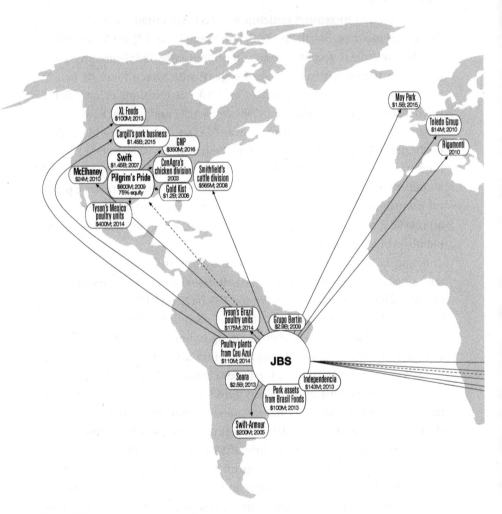

Figure 2.4
JBS: ownership changes, 1996–2016

Corporate Concentration in Global Meat Processing

(BSE) in the United Kingdom, which led to the loss of export markets for U.S. beef (Bell and Ross 2008).

In 2009 JBS acquired a majority stake in the poultry processor Pilgrim's Pride, which had previously acquired ConAgra's chicken division in 2003, and Gold Kist in 2006 (Gold Kist was formerly a producer cooperative that converted to a publicly traded corporation just two years prior). In 2012 JBS increased its stake in Pilgrim's Pride from 68 percent to 75.3 percent,

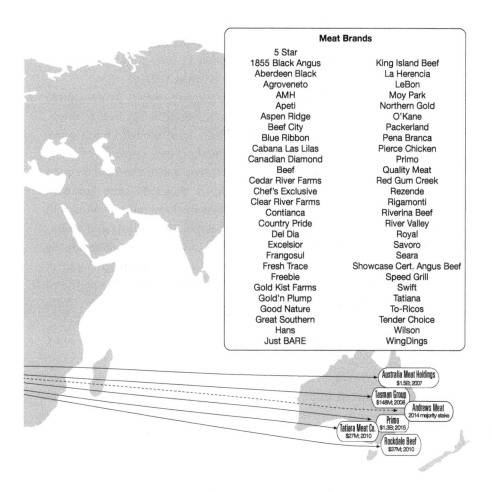

Figure 2.4 (continued)

by paying Bo Pilgrim $107.2 million for his shares in the company (Pilgrim was notorious for handing out $10,000 blank checks on the Texas Senate floor, and for accounting irregularities to reduce tax liabilities) (Richardson 2011).

Just prior to this, in 2008, JBS proposed to acquire National Beef Packing, the fourth largest beef processor in the United States. In a rare antitrust enforcement action, however, it was opposed by the U.S. Department of Justice (USDOJ), due to concerns that the resulting market share would increase prices for consumers and lower prices for ranchers. Because of JBS's more recent legal and political troubles, its BNDES-funded competitor Marfrig saw an opening in 2018, and acquired 51 percent of National Beef Packing for $969 million—a move approved by the USDOJ.

JBS was allowed to make other significant acquisitions in the United States in the past decade, however, including Smithfield's beef business, and Cargill's pork business (for $1.45 billion). Even though Cargill is privately held, and therefore less subject to the short-term demands of investors in publicly held corporations, firm executives decided to focus on grain trading and animal feed production, for which they held more dominant market shares. This move allowed JBS to become even more dominant in meat processing. Tyson made a similar decision in 2014, selling its divisions in Mexico and Brazil to JBS for $575 million, rather than compete directly with a firm that had strong support from a Latin American government. Tyson had acquired three Brazilian firms in 2009, and established joint ventures in Mexico in the 1980s (Constance, Martinez, and Aboites 2010) before ceding these regions to competitors.

Although U.S.-headquartered firms benefited from subsidies that reduce the price of animal feed, as discussed in the case of Tyson, many became vulnerable to takeover in a rapidly consolidating industry in recent decades. Firms headquartered in Brazil have advantages that include government support for the financing of takeovers, but also access to animal feed that is even cheaper than in the United States. Brazil is the largest exporter of soybeans, and second largest exporter of corn, aided by its low cost of production (Sharma and Schlesinger 2017). The price of these commodities is lowered by government supports, such as subsidies for inputs and building/maintaining infrastructure that reduces transportation costs (Fearnside 2001). Corn and soybean prices also exclude many of the negative impacts that result from their production, such as the destruction of rainforests, displacement of small-scale and indigenous farmers, and agricultural chemical pollution. As a result, Brazil continues to rise in the commodity export rankings, recently adding poultry to the world-leading position it already held for beef, along with fourth place for pork exports (Sharma and Schlesinger 2017). Consumption of both beef and chicken is high in Brazil, relative to most other countries (including more industrialized countries), and per capita consumption of all meats in this country is expected to increase further (Zastiral 2014).

Since 2013, JBS has published reports that tout its social and environmental achievements, including claims that it takes measures to ensure suppliers are not destroying rainforests (JBS 2016). It is a member of the Round Table on Responsible Soy, which has standards for sourcing from producers that have not directly converted to soy production. This commitment, however, was only made after months of significant pressure from environmental groups, including Greenpeace. More recently the firm was accused of buying 59,000 cattle from illegally deforested regions of the Amazon (Maisonnave 2017) and is among the leaders of the U.S. Roundtable for Sustainable Beef, which was criticized by a large coalition of nongovernmental organizations (NGOs) for its "blatant greenwashing" (Hamerschlag 2018).

Conclusion: Approaching Limits of Public Acceptance?

Differential government supports have helped shape the winners and losers among meat processing firms. Access to low-cost financing for acquisitions shaped the rapid rise of both JBS and WH Group, which enabled them to take over much larger competitors worldwide, as well as to expand from their initial focus on a single species. This is a risky strategy, requiring high levels of debt, but the increased power that resulted for both firms has helped them to reduce these debts, and plan for even more acquisitions. European-headquartered meat processors, in contrast, have lost power globally because they have made few acquisitions outside Europe (where they also face slow growth in meat consumption and more expensive traceability regulations).

Tyson's ability to remain a global player was shaped by its initial focus on poultry—which was more efficient in conversion of U.S. government-subsidized feed relative to domestic pork and beef packers—and an aggressive strategy of acquiring major competitors. In an increasingly global industry, however, Tyson's focus on the more limited geographic regions of the United States and Asia may result in vulnerability to takeover, much like meat processing firms that have already been acquired by JBS and WH Group in the United States (and the EU and Australia). In the beer industry, for example, Anheuser-Busch was acquired in a hostile takeover by the Brazilian-led/Belgium-headquartered InBev, due to the previous CEO's unwillingness to acquire global competitors (MacIntosh 2011)—this industry is now approaching global domination by just one firm after InBev's acquisition of SABMiller in 2016. Barriers to expanding Tyson's presence in China may play a role in constraining the corporation to slower-growing markets (e.g. more industrialized countries with declining meat consumption), which would disadvantage it even more relative to JBS and WH Group.

The scope of government subsidies provided to Tyson, JBS, and WH Group suggests that without these supports the global meat industry would be far less concentrated in ownership, and the top executives of all of these firms would not be billionaires (nor would executives of firms that were acquired in recent decades have received tens or hundreds of millions of dollars in pay). In addition, the meatification of diets might not have proceeded as extensively, meat production would be less geographically concentrated, and livestock and feed crops would be less genetically uniform. Therefore, the problems posed by these trends, including the increased likelihood of outbreaks of disease in humans, livestock, and feed crops, would not be as substantial as they are now. Government and industry efforts are creating path dependencies or lock-ins for this type of system that will make it more difficult to address these negative consequences by, for example, decentralizing and diversifying production (IPES-Food 2016). Some of these lock-ins include the loss of heritage breeds, the decline of small producers and their knowledge, and the disappearance of smaller processors and their infrastructure.

Government and corporate efforts to legitimize these trends and to conceal their negative impacts are facing greater resistance, however. This has motivated governments to block a few proposed acquisitions or relocations, and even to open criminal investigations in the case of dominant firms in Brazil. In addition, increasing consumer pressure has resulted in promises by the leading firms to phase out some practices that threaten human health (e.g., certain antibiotics, growth promoters like ractopamine) or raise animal welfare concerns (e.g., gestation crates for pork). New communication tools have increased awareness of the problems of a globally concentrating meat industry and enhanced the efficacy of collective action strategies. This is occurring at a time when global climate change and dwindling natural resources, such as fossil fuels and key fertilizers, pose greater challenges to further concentration. The strategies of the largest firms are therefore vulnerable not only to differential support from governments, but also to the biophysical and social limits that are becoming increasingly difficult to overcome, even with substantial subsidies.

3 Aquatic CAFOs: Aquaculture and the Future of Seafood Production

Conner Bailey and Nhuong Tran

Seafood represents an increasingly important source of animal protein available to human consumers. Between 1950 and 2016, the global per capita supply of seafood more than tripled from 6 kg to 20.3 kg (FAO 2018b), more than pork, chicken, or beef (see chapter 1). Seafood also is more important in international trade, with 60 MMT—or 35 percent of total world production—entering export markets (FAO 2018b), compared to 27 MMT for all terrestrially produced meats (chapter 1). If we are going to talk about animal protein sources at a global level, we have to include seafood.

Seafood production comes both from harvests of wild stocks in marine and inland settings as well as from aquaculture. Capture fisheries are dependent upon biologically renewable resources that for the most part are fully exploited. Marine harvests peaked in 1996 and have declined slightly since then (FAO 2018b, 38). Reported harvests from inland capture fisheries have increased steadily since the 1950s, currently accounting for 13 percent of total harvests from all capture fisheries (Welcomme 2011). Aquaculture, in contrast, has emerged as the world's fastest-growing food sector with annual production of 80 MMT in 2016 compared to 1 MMT in 1950 (FAO 2009, 2018b). In 2016 aquaculture accounted for 53 percent of all seafood directly consumed by humans. The primary focus of this chapter will be on aquaculture not only because of its growing importance, but also because aquaculture presents far greater opportunities for corporate investments of the type found in animal agriculture.

In common with the growth of animal agriculture, increased demand for seafood stems from growth in human populations, greater purchasing power of emerging middle-class consumers, and changes in lifestyles and consumer preferences favoring seafood for health reasons. The ability of aquaculture to meet growing demand has been made possible by increasingly intensive production systems that mimic confined animal feeding operations (CAFOs) in animal agriculture. In common with their terrestrial counterparts, aquatic CAFOs can be highly productive and well suited to an industrial approach favoring corporate investors able to mobilize capital on international stock exchanges, including those in New York, Oslo, and Tokyo.

As other chapters in this book have documented (see chapter 2), corporate consolidation in animal agriculture has been driven by public subsidies and taken the form of vertically integrated industries where a handful of corporations control breeding programs, feed supply, processing, distribution, and marketing. Comparable levels of corporate consolidation have not yet taken place within aquaculture or capture fisheries. The model of vertical integration does not fit well with marine capture fisheries, but is more applicable to and is likely to shape the future of aquaculture. Compared to beef, chicken, and pork, however, and despite its recent rapid growth, aquaculture is a relatively new industry and is as yet far removed from oligopolistic control. This is not surprising as the groundwork for domination within animal agriculture was laid in the 1970s, decades before aquaculture began its rapid ascent.

We begin by describing factors behind the rise and continued growth of aquaculture and distinguishing between broad types of production. We will then examine the unintended consequences such as environmental risks and social impacts of aquacultural development. These factors, and in particular issues of environmental and therefore production risk, represent challenges to corporate penetration and domination. We do identify several factors associated with aquaculture that may, nonetheless, lead to concentration of economic power within this sector. One factor is the extraordinarily important role of international trade in seafood compared to animal agriculture, with consolidation of corporations engaged in such trade being one path to consolidation. A second factor is the central importance of feed in intensively managed aquaculture and the potential of corporations with major investments in feed mills to vertically integrate in much the same way as occurred in animal agriculture. A third factor is through corporate control of key breeding stock, including the introduction of genetically engineered fish and shrimp. Just as property rights over genetic materials gave a handful of corporations a dominant role in seed and chemical supply in agriculture, a similar pattern could develop in aquaculture.

The Rise of Aquaculture

Aquaculture has a long history in China and other parts of Asia but is a much more recent development elsewhere in the world. In the United States, for example, commercial production of channel catfish (*Ictalurus punctatus*) was established in the early 1970s (Perez 2006). Commercial production of Atlantic salmon (*Salmo salar*) in Norway started about the same time and was introduced to Chile (the second largest producer of Atlantic salmon) by Norwegian investors only in the 1990s. Commercial shrimp farming in the tropics began in the mid-1970s in Taiwan and expanded into Southeast Asia in the early 1980s, with parallel developments in Ecuador and then other nations of Latin America. Various tilapia species (e.g., *Oreochromis niloticus*) have been part of aquaculture production in Southeast Asia and parts of Africa for many years, but only since 1990 has this species attracted significant commercial investment, shifting tilapia's status from that of peasant food to wide acceptance among consumers of the global north.

The rapid growth of aquaculture starting in the 1990s was made possible by major improvements in hatchery technology, fish genetics, nutrition, and disease management practices. Corporate investments in production facilities, feed mills, distribution networks, and advances in both selective breeding and genetic engineering may open a pathway leading to the expanded corporate domination in aquaculture. Hatcheries are necessary to produce adequate stocking materials in the form of small fish and shrimp to be raised in ponds, cages, or other structures. Fish breeding for many species of fish and shrimp is a relatively new phenomenon compared to plant and animal breeding, but has made rapid progress in part because of the short life spans and high fecundity of most commercially important species. The construction of feed mills has been of particular importance to the growth of aquaculture. Feed represents the highest-cost item in most aquaculture production systems and several of the largest corporations in the seafood industry have feed production as their primary business. The combination of fish and shrimp bred to thrive in congested conditions and intensive feed regimens creates water-quality problems and conditions for bacterial and viral diseases to proliferate. Breeding programs and management protocols have been designed to reduce risks associated with disease but as intensive aquaculture systems proliferate, antibiotic use is on the rise, contributing to the same resistance issues long associated with animal agriculture (Done, Venkatesan, and Halden 2015).

The rapid pace of aquaculture development has been led by producers in Asia, and in particular in China. East Asia, Southeast Asia, and South Asia combined account for 89 percent of total aquaculture production in the world and China alone accounts for 61.5 percent (FAO 2016, 27).[1]

Extensive and Intensive Systems

A key feature of aquacultural development over the past three decades has been the drive to increase intensity of production. Aquaculture systems can be described as ranging along a continuum from extensive to intensive based on stocking density and associated inputs—the higher the stocking density, the greater the need for feed, generally the single-largest variable cost item. Extensive systems are defined by limited or no use of production inputs other than construction of a pond or other enclosure into which wild fish, crustaceans, or mollusks enter, are trapped, and allowed to grow using naturally occurring nutrients. For example, in Southeast Asia brackish water production of shrimp involved opening up floodgates in coastal ponds during the lunar high tides and letting post-larval shrimp and other species such as milkfish (*Chanos chanos*) to enter the pond. Production of oysters and other mollusks may involve little or no use of inputs other than providing a substrate for the capture of juvenile shellfish. Such systems are termed "extensive" because they rely on a relatively large area in a nonintensive manner. Low-intensity "extensive" production systems rely on nutrients naturally occurring in the environment, sometimes supplemented by fertilizers to boost algal growth. A terrestrial equivalent might be grass-fed beef or free-range chickens instead of more intensively managed CAFOs where greater stocking densities combined with intensive feeding and other inputs generate higher yields.

Intensive aquaculture systems depend on supplemental feeds that typically include grains, proteins, essential amino acids, and vitamins. A good example of intensive production is the pangasius catfish (*Pangasius hypophthalmus*) in Vietnam where stocking densities of 40 to 60 fish per m^2 are common, producing fish up to 1.5 kg in a six-month growing cycle and yielding on average 250–300 mt/ha (metric ton/hectare) (Griffiths, Khanh, and Trong 2010). Stocking densities for *Penaeus vannamei*, a widely used shrimp species, commonly are up to 150 per m^2 with harvests of 7 mt/ha in a three-month growing season (Briggs et al. 2004).

Environmental Issues

Dramatic increases in commercial aquaculture production have added appreciably to the global supply of seafood, but not without serious environmental consequences. As with animal agriculture, feed necessary to support intensive aquaculture production systems is globally sourced. The argument that land used to produce animal feeds affects food security where feed grains are grown is applicable for aquaculture as well, but more important is the impact on marine ecosystems associated with production of fishmeal and fish oil. Intensive aquaculture operations also are prone to disease and parasite issues that affect production and can have wider ecosystem impacts. Escapes of exotic species from pens and ponds also pose threats to aquatic ecosystems. All of these issues will be discussed.

Marine Ecosystem Impacts

Each species has different feed requirements, but in broad terms we can speak of carnivores like salmon, shrimp, and tuna, and omnivores like tilapias and carps. The carnivores (or more specifically piscivores as they require protein from seafood) depend heavily on fishmeal and fish oil provided primarily from marine capture fisheries (with additional input from seafood processing waste[2]). Approximately 25 percent of all marine harvests from the wild (20.9 MMT) are small fish (e.g., anchoveta, menhaden, caplin) used for production of fishmeal and fish oil and not consumed directly by humans (table 3.1). Piscivores require feeds with 18 percent to 30 percent by volume of fishmeal as well as fish oil for optimal growth and flesh quality. Omnivores such as tilapia, carp, and pangasius catfish often are fed a diet containing small amounts of fishmeal supplemented by protein from plant sources, typically soybeans.

Demand for both fishmeal and fish oil is likely to continue increasing because the piscivores are high-value commodities particularly prized among consumers in industrialized nations of the global north. The supply of fishmeal and fish oil has long been recognized as a limiting factor in

Table 3.1

Global capture fisheries and aquaculture production and utilization, million metric tons, 2009–2016

	2009	2010	2011	2012	2013	2014	2015	2016
Capture								
Inland	10.5	11.3	10.7	11.2	11.2	11.3	11.4	11.6
Marine	79.7	77.9	81.5	78.4	79.4	79.9	81.2	79.3
Total capture	90.2	89.1	92.2	89.5	90.6	91.2	92.7	90.9
Aquaculture								
Inland	34.3	36.9	38.6	42.0	44.8	46.9	48.6	51.4
Marine	21.4	22.1	23.2	24.4	25.4	26.8	27.5	28.7
Total aquaculture	55.7	59.0	61.8	66.4	70.2	73.7	76.1	80.0
TOTAL	145.9	148.1	154.0	156.0	160.7	164.9	168.7	170.9
Human consumption	123.8	128.1	130.0	136.4	140.1	144.8	148.4	151.2
Non-food uses	22.0	20.0	24.0	19.6	20.6	20.0	20.3	19.7
Population (billions)	6.8	6.9	7.0	7.1	7.2	7.3	7.3	7.4
Per capital food fish supply (kg)	18.1	18.5	18.5	19.2	19.5	19.9	20.2	20.3

Source: FAO 2016, 2018b.

the expansion of aquaculture (Naylor et al. 2000; Tacon and Metian 2009). Research to improve feed conversion ratios has achieved significant success, but aquaculture feeds still utilize 68 percent of global supplies of fishmeal and 88 percent of global fish oil (Naylor et al. 2009). Demand for aquaculture feeds is likely to continue growing, putting increased pressure on small fish species that play a key role as forage fish in marine ecosystems supporting bird life, other fish, and human consumers (generally the poor in developing countries) who depend on small fish for food.

Disease and Parasites

Intensive production systems made possible through supplemental feeding offer the prospect of higher yields per hectare but they also pose higher risks. Disease is a constant threat in densely stocked ponds or cages due in part to crowding and stress, and caused also by water quality problems related to the production process itself (uneaten food and fecal materials, polluted water discharged into the environment by other aquaculture operations or other industries in the watershed). Controlling a disease outbreak in an aquatic ecosystem is difficult and disease organisms often spread rapidly across wide areas and even from one part of the world to another. For example, a virus causing infectious salmon anemia simultaneously affected Atlantic salmon produced in Norway, Scotland, Canada, and Chile. Salmon producers in Chile have been using large quantities of antibiotics, including 1.2 million pounds in 2014 (Esposito 2016); antibiotic use in aquaculture is a common practice in many parts of the world. Wild salmon in the Pacific Northwest have been affected by sea lice when they swim near caged salmon infested with this parasite (Krkošek et al. 2006). Disease outbreaks have been a constant problem associated with intensive production systems used by shrimp producers in Asia and Latin America. Channel catfish viral disease in the United States is a problem related to stress induced by high stocking densities in ponds.

Salmon farming also imposes a number of externalities, the severity of which will depend on physical conditions and how the production process is regulated and managed. Salmon are raised in large pens typically in protected seawaters such as fjords. Fecal waste and uneaten food pass through the pens and the water column to the ocean floor. If the waters are deep, there is thorough tidal flushing, and if the number of pens is appropriately limited, these wastes are easily assimilated through natural processes and the likelihood of diseases greatly reduced. Where these conditions do not exist, disease problems and the use of antibiotics become common.

Escapes

Another risk associated with farmed salmon involves escapes into the wild. Escapes are common both as small "trickles" of fish and in more catastrophic escapes. In August 2017, approximately 120,000 Atlantic salmon (*Salmo salar*) escaped from pens in Puget Sound, British Columbia (Mapes 2017). In this setting, Atlantic salmon are an exotic species, one that has established breeding populations in three streams in British Colombia, representing a potential threat to several species of Pacific salmon that are under heavy fishing pressure at sea and habitat loss on land (Thorstad et al. 2008). Escapes of Atlantic salmon from pens in Europe, where the species is native, also can be disruptive because the escapees have been selectively bred for adaptability to pen culture and willingness to feed on pellets. When domesticated and wild salmon interbreed in the wild, the resulting progeny have lowered fitness and reduced lifetime success (Thorstad et al. 2008).

Escapes of exotic species can cause havoc on aquatic ecosystems. Water is a highly connective medium and once a fish is released into the environment there is a real danger of it traveling along and across watersheds. Flooding of aquaculture operations in the early 1990s led to the escape of silver and bighead carp into the Mississippi River in the United States. These carp species have moved up the river, displacing native fish species. There is concern that these fish will make their way into the Great Lakes through the Chicago Ship and Sanitary Canal, which connects the Mississippi River to the Great Lakes. In 2017, an adult carp was found within nine miles of Lake Michigan, eighteen miles past an electronic barrier created to stop their spread. There is great concern that native species will be displaced if carp make their way into the Great Lakes (Alliance for the Great Lakes 2017).

Societal Dislocations

Environmental and societal disruptions caused by the introduction of intensive aquaculture systems can be separated analytically, but as the following example of shrimp farming demonstrates, social and environmental costs are two sides of the same coin. Mangrove destruction undermined local livelihood strategies and food security. In this section we focus on resource conflicts and food security as examples of societal dislocations that can be caused by the introduction of intensive aquaculture.

Resource Conflicts

Land and water used in aquacultural development did not suddenly appear from a big basket of unused resources. Rather, these resources often were used for other purposes, whether for agriculture or for ecosystem services. That aquacultural development can impose social and environmental costs has been known for at least thirty years, when the first critique of shrimp farming was published in the peer-reviewed academic literature (Bailey 1988). Central to this critique was the impact of rapidly expanding shrimp farms on millions of hectares of mangrove forests that were destroyed

(Thomas et al. 2017). Most investors were outsiders, urban elites who used their financial resources and political connections to gain access to what were generally regarded as public lands. Coastal fishing communities depended on mangrove resources for a wide range of products, including building materials and fuel wood. Mangrove forests were particularly valuable as a nutrient-rich habitat where many commercially important finfish and crustaceans found food and shelter as juveniles before making their way to the open ocean. Mangrove destruction undermined the resource base that coastal communities depended upon for subsistence and non-subsistence needs. Where small-scale extensive aquaculture was practiced in coastal areas, these producers were bought out or forced out by disease outbreaks associated by wastewater from shrimp farmers using intensive methods of production.

In more recent years, many shrimp farmers have learned that soils associated with mangrove often are unsuitable for sustained production of shrimp (due to acidic soil conditions) and have shifted operations to lands just outside the mangrove forests. In Asia, many of these lands previously were used for rice production. Intensive production of shrimp requires large volumes of fresh water to maintain optimal salinities as pond water evaporates. Coastal aquifers are vulnerable to over-exploitation and a common problem faced by coastal communities where shrimp farms are present is saline intrusion into freshwater aquifers. Where this happens, coastal residents may be forced to purchase fresh water for household purposes.

Resource conflicts between aquaculture and other industries are not uncommon. Industrial, urban, or agricultural land uses upstream can affect water quality and threaten or foreclose opportunities for aquaculture. Rice-fish culture, a low-intensity form of aquaculture practiced in Asia for millennia, has greatly diminished as farmers adopted herbicides and insecticides as part of more intensive rice production systems. Shellfish production effectively carves out a part of the near-shore coast, turning public access waters used by fishers and recreational users into private property. Because such shellfish are likely to be placed in the most productive waters, local fishers in particular might feel that their livelihoods are threatened by the spread of aquaculture. In extreme cases, aquaculturists have blocked off direct access to fishing grounds, making it necessary for fishers to travel extra distance each day.

Food Security

As is the case with animal agriculture, expanded production of protein does not necessarily translate into food security for all. There should be no surprise to this given the nature of markets in a global capitalist system where product flows to markets able to pay the most.

The intensification of aquaculture has contributed significantly to the growth of international trade in seafood, which in 2016 was valued at US$142 billion (compared to US$71.9 billion in 2004). The growing importance of international seafood trade has created new space for transnational corporations to expand their operations; most of the top corporations in the seafood industry are vertically integrated producers, processors, and distributors of seafood based in Japan and Europe (Österblom et al. 2015). Vietnam and Thailand alone accounted for 10 percent of global exports and four of the top seafood exporters in 2014 were non-industrialized nations of the global south (FAO 2016). Broadly speaking, we can describe a pattern of high-quality protein in the form of seafood being exported from the global south to the global north where consumer buying power creates the most profitable markets (Kagawa and Bailey 2006).

There are important caveats to this broad statement. Belton, Bush and Little (2018) argue that this critique misses the important role of commercial producers, particularly in Asia, who produce for poor and middle-income domestic consumers and who therefore make important contributions to food security. They make the important point that in many countries, over 90 percent of all aquaculture production is consumed domestically, though there are other countries (e.g., Thailand, Vietnam, etc.) where that figure is one-third or less of all production. Asche et al. (2015) argue that seafood exports generate income and employment in the global south and that seafood imports from the global north to south, which take the form of lower-valued products, balance out the net nutritional balance. These authors note in conclusion, however, that aggregate trade figures tell us nothing about the impact of seafood trade on the poorest of people in developing countries, only that international trade provides the means by which developing countries are able to improve societal welfare.

International trade in seafood from the global south to the global north involves more than the shipment of high-quality protein; it also represents allocation of water, land, and other resources to that end. Intensive production systems for shrimp, for example, have disrupted coastal ecosystems and undermined food security in many coastal communities in South Asia, Southeast Asia, and Latin America. The impact comes from not only the export of protein but also the transformation of ecosystems and the livelihood strategies associated with those ecosystems.

The Green Revolution of the 1970s led to dramatic increases in food production, but benefits did not always reach the landless or the poor. So too the growth of aquaculture has increased seafood supplies globally, an important point not to be missed. As with the Green Revolution, or any other major technological change, there are likely to be winners and losers. Consumers who can afford to buy fish, whether in developing or developed nations, are the beneficiaries of aquacultural development. Rural people displaced by aquacultural development may find employment in the ponds or processing facilities or other sectors of the economy. But we also can predict that there are those who will be left behind and who will experience increased food insecurity even as fish production increases simply because they cannot access the market.

Constraints to Corporate Penetration and Consolidation

There are multibillion-dollar corporations engaged in vertical integration involving some combination of marine fisheries, feed production, and aquaculture (as will be explained), but nothing exists remotely resembling the kind of oligopolistic structure found in pork, chicken, and beef described elsewhere in this volume. One reason for this is that aquaculture is simply a newer industrial actor. Animal agriculture had industrialized by the 1970s and the current corporate consolidation is simply the most recent phase of that process. Aquaculture, in contrast, only became a major source of production in the last three decades. Additionally, aquaculture remains a relatively risky enterprise compared to animal agriculture. This may change with research and experience, but problems of disease and parasites are more difficult to control in an aquatic environment than in a terrestrial one where animals can plainly be seen and air quality can quickly be improved with ventilation.

Perhaps a larger constraint to consolidation of corporate power has to do with the highly fractured nature of aquaculture in scale, technologies, and species, not to mention that the product—seafood in its many permutations—is also supplied by a largely separate industry that depends on marine harvesting of aquatic life-forms from the wild. Seafood production systems are highly diverse and may resist the kind of corporate consolidation found in animal agriculture. Because most marine fishery resources are effectively open access, the ability of corporations to dominate the marine sector is limited. There are niches where corporate investments rule the waves. Open ocean or distant water fisheries are the natural realm of corporate actors who invest in large factory ships that both catch and process fish into frozen fillets or other products. But the global fishing fleet in 2014 was made up of 4.6 million fishing vessels, more than one-third of which did not have engines. Of motorized craft, 85 percent were less than 12 meters in length (FAO 2018b, 35). Only 2 percent of all fishing vessels were longer than 24 meters in length. These larger vessels doubtlessly landed a disproportionate share of the total catch, but even so, the marine fisheries sector can be characterized as a sea of small boats serving numerous markets. Österblom et al. (2015) found that the 13 largest corporations in the world involved in marine capture fisheries accounted for 11–16 percent of global harvests. This is a significant fraction but far from the corporate concentration found in beef, chicken, and pork.

Aquaculture production systems also are highly diverse, with the FAO (2016) reporting 369 finfish, 109 mollusks, and 64 crustaceans being grown in fresh water, brackish water, and coastal marine settings from the arctic to the equator. This wide array of species is produced in earthen ponds, above-ground tanks, raceways, cages, and pens in freshwater, brackish water, and seawater. Production systems range from small backyard ponds where fish are fed kitchen scraps to intensively managed ponds with high stocking densities and feeding regimes. Some systems are built to be entirely self-contained, with filtration systems removing waste products before water is recycled into the tank or pond; in other systems wastes are absorbed into the larger ecosystem. Where use of inputs is limited, this presents few problems but where high stocking and feeding rates are used, eutrophication, disease outbreaks, and fish kills through oxygen depletion are serious risks. Managing such risks is an important part of intensive production systems of the type that would attract corporate investors.

Pathways for Corporate Penetration and Consolidation

Compared to capture fisheries, opportunities for corporate penetration and domination in aquaculture are greater because key factors of production are more easily controlled. In contrast with the open-access nature of most wild fisheries, property rights can be established over land and water used for aquaculture production either through direct ownership or through long-term leases. These property rights in turn enable investments in ponds, raceways, or other physical infrastructure. Aquaculture producers also are able to utilize key inputs including stocking materials and feeds to gain higher yields. Investments to maintain adequate water quality through pumping or aerators or both also can be employed to increase the intensity of production. In short, aquaculturists have a high degree of control over the production process.

We see three pathways through which consolidated corporate power in aquaculture could emerge. The first has to do with control over international trade, the second with vertical integration centered on domination of feeds, and the third through control of genetic materials. These pathways are analytically separable but are likely to be pursued simultaneously by corporations seeking to establish a dominant role in aquaculture.

International Trade

One path to corporate consolidation in aquaculture is through control of international trade, which represents 35 percent (by value) of global seafood production, capture and culture combined (FAO 2018b). Production systems for wild harvests and aquaculture are very different but once seafood enters the market, differences between captured and cultured seafood disappear, not only from the perspective of individual consumers but more importantly because corporations that trade in seafood handle product from both sources.

The high value in international trade is central to understanding the profitability of corporations in the seafood industry. The United States and Japan were the leading importers, together accounting for one-quarter of global seafood imports in 2016 (FAO 2018b, 55); the United States ran a seafood deficit of US$14.7 billion in 2016. China and Norway were the world's largest exporters, together shipping almost $US31 billion worth of seafood around the world. Fish exports from the global south made up 54 percent of total value and 59 percent in total weight in 2016 with a net trade balance of $37 billion, higher than tobacco, rice, sugar, and other types of meat combined (FAO 2018b, 57). Unlike trade in other types of meat, where trade flows generally are from the global north to the global south (chapter

1), the flow of seafood moves primarily from the global south to nations of the global north.

The largest corporations in the seafood industry are the large trading houses of Japan. Maraha Nichiro has operations in 65 countries while Nippon Suisan Kaisha and Kyokuyo operate in 32 and 15 countries, respectively (Österblom et al. 2015). The top one hundred seafood corporations had combined revenues of US$93 billion in 2015. The ten largest of these all had revenues over US$2 billion and accounted for more than one-third of total revenues (Undercurrent News 2016). These are large corporations but still small in comparison with Tyson Foods (n.d.), with sales of US$38.3 billion in 2017.

Vertical Integration around Feed Mills

Several corporations with origins in seafood trade are expanding investments into feed mills, seafood processing, and aquaculture production facilities. As is true for animal agriculture, intensification of aquaculture production is highly dependent upon feed mills which use many of the same inputs as found in other animal feeds, but with a higher concentration of fishmeal and fish oil derived primarily from wild stocks of small pelagic fish such as anchoveta, menhaden, and capelin. Fishmeal and fish oil are particularly important elements in feeds for carnivorous species such as salmon and shrimp and also are used in small quantities for feeds formulated for noncarnivorous species such as tilapia and carp. Corporations involved in seafood trade are well placed to access raw materials used to produce aquaculture feeds, and even more so if they also are involved in seafood processing, since wastes from such facilities are the second most important source of fishmeal and fish oil.

The importance of feed in aquaculture is highlighted by FAO (2018b, 22) data showing that the percentage of cultured species provided supplemental feeds versus those that are not increased from 45 to 70 percent between 2001 and 2016. On the supply side, six of the world's top 12 seafood corporations identified by Österblom et al. (2015) were involved in feed production. Four of these six are integrated either in the harvest of small pelagic fish for fishmeal and fish oil (Austevoll Seafood of Norway and Pacific Andes of China) used in the production of feeds or are involved in aquaculture production (Dongwon Group of South Korea and Charoen Pokhand of Thailand). Belief that aquaculture is an attractive investment in

2015 led Cargill to purchase the Norwegian feed manufacturer EWOS. The Dutch holding company SHV purchased Nutreco, owner of Skretting of Norway. Corporate buyouts are of course one strategy through which large corporations like Cargill establish a position in a market in order to grow their business. Feed is the single largest cost item in aquaculture production and feed suppliers are in a position to offer other products and services to producers. What has not yet emerged in aquaculture is the type of vertical integration seen in hog and poultry production whereby a producer buys stocking materials and feed from a corporation and then sells the harvest back to that corporation for subsequent processing, distribution, and marketing. However, the potential exists for this to occur.

From Selective Breeding to Genetic Engineering
The pace of selective breeding has accelerated in the past several decades due to scientific advances in genetic and molecular biology. Because of their high fecundity (ability to produce a large number of eggs) and short lifespans, fish are excellent candidates for research on genetic improvement. As an example, Norwegian salmon farmers began to experiment with selective breeding in the early 1970s, looking for fish that tolerated crowded conditions, fed well, were disease resistant, grew rapidly, and had the right flesh color and fat content. Gjøen and Bentsen (1997) reported that after four generations, the growth of selected salmon to smolting (the life stage where salmon adapt to seawater) was twice as fast as for wild salmon.

Advances in genetic and molecular biology have provided a scientific basis for improving the genetic composition of farmed fish whether through conventional selective breeding, interspecies hybridization, or genetic engineering. Each of these technologies creates opportunities for more intensive production practices in industrial CAFO settings. More importantly, a pathway to corporate power is created by controlling the genetic material that provides for faster growth, disease tolerance, and other desirable traits.

The ability to control sex is an important and widely used tool in aquaculture. Energy used to produce sexual organs, sperm, or eggs can be harnessed to increase growth rates and often to improve flesh quality. Male channel catfish and tilapia grow faster than females, making production of a single-sex male population an important management goal. The sexual characteristics of fish and other aquatic organisms can be manipulated relatively easily through use of hormone treatments or changes in pressure or temperature at certain stages in the growth cycle.

The relatively plastic genetic makeup of fish allows for genetic improvement through polyploidy. Most fish species are diploid, meaning that they contain two sets of chromosome pairs, one inherited through the mother and one through the father. Triploid fish can be created through abrupt pressure and/or temperature changes or through chemical exposure. Triploids have a third set of chromosomes representing additional genetic material in each cell and as a consequence, each cell is somewhat larger (as is the whole fish). The key advantage of triploids is that they are sterile but they also tend to grow more quickly because energy is directed to body growth rather than growth of sex organs. Triploidy often is promoted where a high proportion of an animal's energy goes into reproduction. Triploidy represents a step beyond simple mono-sexing and a further step in the path toward intensification.

Hybridization can involve breeding programs involving individuals of the same species but from different strains that have different genetic backgrounds. Such hybridization results in greater genetic variation in the offspring and the potential for hybrid vigor. In the United States, crossing of the channel catfish with the blue catfish has produced a hybrid that grows 25 percent more quickly than either channel or blue catfish, tolerates lower oxygen levels, has a better fillet percentage, and is more resistant to disease. The production of hybrids opens up commercial opportunities for breeders and increases potential for more intensive CAFO operations.

Mono-sexing, triploidy, and hybridization are now standard approaches to improving the genetic resources available to aquaculture producers. More recently, major advances are being made in mapping the entire genome of several commercially important aquacultural species including salmon, catfish, cod, rainbow trout, tilapia, carp, striped bass, shrimp, oyster, and scallop. Such mapping allows researchers to know where genes controlling desirable traits are located, allowing for highly efficient and targeted selective breeding for such traits as disease resistance.

Investments in selective breeding programs, genome mapping, and development of hybrids have economic value as a form of intellectual property over which certain rights are protected either by contract or by patent. Research on hybrid catfish at Auburn University resulted in a license to Eagle Aquaculture, Inc. to develop this technology. In Europe, major corporations including Wessjohan, Landcatch, and Stofnfiskur have begun to exert a stronger proprietary interest in the genetic resources that they control. The German EW Group (a leader in poultry genetics through its subsidiary Aviagen) has gained majority ownership of AquaGen AS, a Norwegian company with 35 percent of the world market on salmon (Rosendal, Oleen, and Tvedt 2013).

The next step in opening space for corporate growth within aquaculture is likely to be through genetic engineering (GE). GE research on fish is underway in China, Cuba, India, Korea, the Philippines, Thailand, and the United States (Fu et al. 2005). The U.S. Food and Drug Administration (FDA) approved the first transgenic animal, a genetically engineered Atlantic salmon, for human consumption in 2015 and Health Canada issued its own approval in 2016 (U.S. FDA 2015; Health Canada 2016). AquaBounty Technologies (ABT) of Massachusetts is a small company that spent almost two decades developing and obtaining FDA approval for this transgenic Atlantic salmon that they claim can reduce the time needed to reach marketable size from thirty to eighteen months.

Future Directions

Aquaculture as an industry is a relatively new actor on the global meat world stage. Nonetheless, scientific, technical, and organizational changes that have shaped contemporary animal agriculture, described elsewhere in this volume, provide the basis for anticipating the future of seafood production.

The rise of aquaculture and growth in international seafood trade are closely linked and there are a number of transnational corporations that have built their fortunes in the larger seafood industry, with investments in some combination of fishing vessels, aquaculture facilities, feed mills, and international trade. In their study of 160 corporations in the seafood business, Österblom et al. (2015, 4–5) found that the top 10 percent accounted for 38 percent of total revenues and were especially dominant in production of feeds for salmon and shrimp (68 percent and 35 percent of global totals, respectively). They likened these corporations to "keystone species in ecological communities in that they have a profound and disproportionate effect on communities and ecosystems and determine their structure and function to a much larger degree than what would be expected from their abundance" (Österblom et al. 2015, 1). In other words, a small number of large corporations in the seafood industry have a strong influence on specific segments of the industry.

Property rights over seed technologies have led to a concentration of corporate power in agriculture (Kloppenburg 2004) and the ability to patent transgenic fish opens the door to similar opportunities. If ABT's transgenic salmon proves a commercial success, other transgenic species will enter the market. If GE provides for improved growth rates and other production advantages, not only will transgenic fish gain a share of the seafood market but also corporations able to patent these fish will have established key positions within the industry. Through the process of corporate merger, corporate innovators such as ABT might be purchased by larger corporations once these potential buyers are convinced that the technologies have merit.

At present, corporate investments in aquaculture are highly concentrated on a small number of high-value species, particularly shrimp and salmon, which serve markets of the global north. These high-value species are the ones that are attracting corporate investments in selective breeding and genetic engineering because that is where financial gains are most readily realized. The same is true with investments in feed mills, which are likely to be placed near production sites for species with the greatest international markets, including shrimp, salmon, tilapia, and pangasius catfish. These investments in seed and feed will result in increased corporate power among producers and processors of a limited set of important species.

There remains the question of how corporate interests will interact with the millions of limited resource aquaculturists around the world with small ponds or cages, few financial resources, and limited technical expertise. Seventy percent of all aquacultural production comes from producers using supplemental feeding. Included in this total are many small-scale producers. Importantly, 30 percent of all aquaculture production does not include supplemental feeds, representing a potential growth market. Small-scale producers may not be tied to global markets but increasingly they are being influenced by transnational corporations that supply feed, pharmaceuticals, and other inputs to support higher productivity. In the absence of effective extension programs in most parts of the developing world, such input suppliers often provide, directly or through their sales agents, technical advice and credit to all types of producers eager to increase their production through increasing stocking density and input use. In this way, even where corporations are not directly engaged in production and where the product does not enter global markets, the influence of corporate actors can be expected to penetrate deeply into many parts of the world.

The likelihood that aquaculture might follow the path of chicken production, with a small number of integrators dealing with individual producers, is a realistic prospect in the decades to come for a limited number of high-value species. The aquaculture sector is likely to develop a relatively small number of well-capitalized intensive production systems serving urban markets of the industrial north and the newly prosperous elites of emerging nations. At the other end of the spectrum will be limited resource producers who use few if any purchased inputs to supply local markets. In between there will be a diverse mix of producers serving local, regional, or even national markets and operating somewhere on the continuum between intensive and extensive production systems.

Corporate consolidation in the aquaculture sector may be further advanced through certification efforts, pushed by environmental, non-profit organizations and embraced by major retailers in the global north, designed to encourage environmentally sound and socially just production of seafood. Conservation and consumer activist groups have built on the experience of the Forest Stewardship Council and the Marine Stewardship Council to develop the Aquaculture Stewardship Council, which will certify seafood products. The Global Aquaculture Alliance, an organization of corporate actors within the seafood industry, has established its own Best Aquaculture Practices certification program. Certification programs involve third-party evaluations and cost more money than most small-scale producers in the tropics can afford. The difficulty in assuring chain of custody of seafood from numerous small-scale producers through middlemen and on to processors, exporters, and retailers is a serious obstacle to participating in certification programs (Tran et al. 2013). Because major retailers like Wal-Mart are moving towards limiting the sale of seafood to certified products, in response to consumer demand, the ability to participate in certification programs will dictate the ability to access the most lucrative markets. Corporate producers will have the advantage of scale and technical expertise to meet certification standards and be able to afford costs of certification.

In sum, we expect corporations in the seafood industry generally, and the aquaculture sector in particular, to play an increasingly important role. Some small-scale producers will either be eliminated or incorporated into vertically-integrated commodity systems as contract growers. We have already seen corporate mergers and movements toward vertical integration in this sector. Technological changes, including genetic engineering, will contribute to further consolidation in aquaculture production. And finally, certification systems may provide important advantages to corporate producers compared to small-scale producers in gaining access to the most lucrative of markets.

Notes

1. There have been disputes regarding the accuracy of fisheries data from China, particularly regarding marine capture fisheries (Watson and Pauly 2001) and there is some uncertainty regarding accuracy of FAO statistics on some forms of aquaculture production, notably from China (Campbell and Pauly 2013; FAO n.d.).

2. Waste by-products from pangasius catfish are now used as inputs for fishmeal production in Vietnam.

II From Global to Local

The global forces propelling the expansion of the meat industry have, on the one hand, increased the world food supply in the sense that there is more meat. On the other hand, however, this increased food supply has not erased hunger and malnourishment, as 821 million people still suffer from hunger (FAO 2018a, 3, fig. 1).[1] Indeed, we argue that the global meat industry contributes to increased vulnerability to food insecurity globally, not to mention increased health problems for approximately another one billion people (see Otero et al. 2018). While the global expansion of the meat industry has increased the supply of food available to some people, it has undermined food security for many in at least three important ways: (1) diverting production in many regions toward profitable exports, (2) reducing access to land in many parts of the world, and (3) reducing wages for workers within the meat industry.

First, the global expansion of meat production goes hand in hand with increased international trade. We have already seen how much meat production and trade have increased over the past twenty-five years. As noted in chapter 1, the global meat trade expanded rapidly from 1998 to 2015, more than doubling from 12.6 MMT to 27 MMT in less than two decades. During this same period, the percentage of global meat production exported also increased, from 6.9 percent in 1998 to 10.5 percent in 2015.[2] In some instances, the increase in meat exports came in countries where food insecurity was a significant problem. India, for example, is a country that has long faced a chronic problem of food insecurity, with about 17 percent of the population categorized as suffering from hunger between 2011 and 2013. Despite such food insecurity, India was the world's leading beef exporter in 2014, having exported 2 MMT of beef, which surpassed the exports of Australia, Brazil, or the United States. Following the global food crisis in 2008, both beef production and exports in India increased substantially. Given the increase in exports, in particular, this beef production was not aimed at alleviating hunger in India. Furthermore, the focus on meat production and exports can divert resources, such as land and water, away from the production of food accessible to poorer segments of society (Winders 2017, 101–104).

Second, the global expansion of meat production has also reduced access to land in many parts of the world and inhibited subsistence production in many regions as more land has gone into pasture and feed grain production. This has amounted to a process of dispossession in which peasants and smallholders have lost rights and access to land. Part of the process of the "global land grab" has involved attempts to acquire large tracts of land to accommodate greater meat production (Lavers 2012). In other instances, corn and soybean production has spread to cover more land (Sauer and Leite 2012). Such land grabs are also effectively "water grabs" because expanding meat, corn, and soybean production requires greater amounts of fresh water, which in some regions means a diversion of fresh water from local populations (Rulli, Saviori, and Odorico 2013).

Finally, the global expansion of the meat industry has, at least in part, involved a search for lower production costs. This includes attempts to lower wages for workers in the meat industry. Consequently, many workers in the meat industry, itself, may find themselves in precarious financial positions. In the United States and elsewhere, large corporations in the meat industry, such as Tyson, have worked to squeeze more production out of workers while also developing strategies to reduce workers' pay (Striffler 2005). This process has included attempts to reduce union representation in the meat processing industry and shifts in the composition of workers, such as relying more heavily on immigrants. In the middle of the twentieth century in the United States, meat processing was unionized and meatpacking workers had won middle-class wages, good benefits, and better working conditions (Stull and Broadway 2004). By the 1980s, however, corporations had developed a number of effective strategies in the struggle with workers and unions, including corporate consolidation, the relocation of processing plants away from urban areas, and a greater reliance on hiring immigrants (Ribas 2015; Striffler 2005). These strategies weakened workers in terms of their ability to demand higher wages, better benefits, and safer working conditions. One result of this kind of class conflict, where the balance of power has shifted toward large corporations, has been declining wages and job stability for workers. This, in turn, increases the likelihood of food-insecure households for those working in the meat processing industry.

In addition to contributing to food insecurity, the growth of global meat production has also contributed to negative health consequences for many people in at least three ways. First, there is the role that meat plays in the so-called "neoliberal diet," which has involved increased consumption of sugar, oils, and processed foods and at the same time less fresh fruits, whole grains, and legumes (Otero 2018; Otero et al. 2018). With the oversized growth of meat production, we have seen a dramatic increase in meat consumption, but the types of meat consumed tend to take on class dimensions, with poorer people (including ironically those that work in meat processing plants) consuming more processed and cheaper cuts of meats that are generally higher in fat, salt, or other food additives. In the global north, this has contributed to the obesity-hunger paradox, whereby poorer people are more likely to be obese due to their consumption of more highly processed foods that are generally cheaper and more accessible than fresh foods, including lean cuts of meat (Carolan 2013).

Second, global trade in meat often involves different parts of the animal being shipped around the globe to different locations, with less desirable types of meat often landing in the global south. Gewertz and Errington (2010, 1) offer a good example of this in their discussion of the meat trade in the south Pacific: "Our story is about a fatty, cheap meat eaten by peoples in the Pacific Islands, who are among the most overweight in the world. Lamb or mutton flaps—sheep bellies—are often 50 percent fat. They move from First World pastures and pens in New Zealand and Australia, where white people rarely eat them, to Third World pots and plates in the Pacific Islands, where brown people frequently eat them—and in large amounts."

A similar tale exists for many countries in the global south, including South Africa, as mentioned in chapter 6 of this volume. While this volume largely focuses on the growth of global meat production, we would be remiss if we did not highlight the dual paradox of the growth of global meat contributing to *both* food insecurity and negative health consequences for people around the world.

The chapters in part II highlight some of these dynamics. Schneider, in chapter 4, reveals the ways in which Chinese farmers either become more industrial in their own pork production or risk dispossession from their own land. Rudel, in chapter 5, demonstrates how trade liberalization and international trade in meat contributed to the unintended consequence that some farms (generally, owner-occupied) have silvopastoral landscapes that create lower emissions for cattle ranching. Finally, in chapter 6, Freshour shows how corporations in the poultry industry have shifted their hiring practices to increase their control over workers and reduce wages.

Notes

1. This is an increase over the estimated 784 million people suffering from hunger in 2015 and 804 million people in 2016 (FAO 2018a, 3, fig. 1). The increase in world hunger over the past few years is a departure from the long-term decline found in previous years. Some scholars have challenged the accuracy of this long-term decline, which rests on changes in who is categorized as suffering from hunger (Lappe et al. 2013).

2. Calculated using data from USDA FAS n.d.

4 China's Global Meat Industry: The World-Shaking Power of Industrializing Pigs and Pork in China's Reform Era

Mindi Schneider

Nineteen-seventy-nine was a watershed year in the global meat industry. After growing in tandem for decades,[1] pork overtook beef, initiating the meteoric rise of "the other white meat" to the top of global meat markets.[2] Underlying the pork boom was China, or more specifically, the reforms to China's economy that began with Reform and Opening (*gaige kaifang*) in 1978. By liberalizing selected agricultural markets, supporting the emergence of private and state-owned agribusiness firms, and encouraging scaled livestock and feed operations, state policies were instrumental in the production upswings that brought more pork to the Chinese people, and pushed pork to dominance in the global arena. Pork's rise was China's rise.

Throughout China's reform era (post-1978), policies and markets have further industrialized and commercialized livestock and pork production. The Household Responsibility System (HRS) in 1981 decollectivized the countryside, spurring privatization and private entrepreneurship, and prompting new forms of commercial and contract farming. Liberalization of soybean imports from the early 1990s transformed the pork industry, freeing livestock feed needs from the constraints of China's high-population/limited-land conundrum. And the definition and adoption of agricultural "modernization" in development policy and practice since the 1990s further restructured agricultural production and rural social relations in ways that resulted in massive production increases, along with social and dietary change, and environmental degradation. Taken together, these transformations have propelled China's pigs and pork to become world-shaking things. By 1995, farmers and companies in China were producing 40 percent of all the pork in the world.[3] Today, although production growth has stagnated since 2015, China remains home to almost half of the world's pigs, half of the world's pork production, and half of the world's pork consumption.[4] China produces five times as much pork as the United States, and twice the amount in the European Union (figure 4.1).

While pork is not the only meat rising in China, it is certainly the star of the show. Overall per capita meat consumption has quadrupled since 1980; by 2013, the average Chinese person ate 65 kg of meat.[5] In 1980, pork accounted for 82 percent of China's meat consumption, declining to 61 percent in 2013, as chicken has become increasingly important in terms of production output and consumption levels.[6] Still, pork remains the heart of the reform-era meat boom (figure 4.2).

These relatively rapid changes beg several questions: Why pork? Why now? How did the pork boom happen? With what consequences? And where is it heading in the future? This chapter offers some answers. It highlights cultural histories and meanings around pigs, pork, and peasants; the role of the state and corporations in the reform-era pork industry; and social and environmental implications of the pork boom.[7] The chapter looks at some of the ways in which transforming pork production and consumption has been premised on the industrialization and capitalization of the swine sector *within* China, while considering relations *beyond* China as

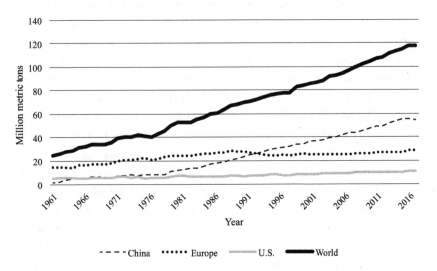

Figure 4.1
World pork production, 1961–2016
Source: FAO 2019.

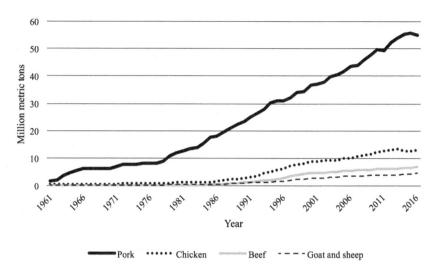

Figure 4.2
Meat production in China, 1961–2016
Source: FAO 2019.

well. It argues that China's is a global meat industry: resource flows, production, political economic relations, and political ecological consequences transcend borders, and leading Chinese pork firms are increasingly transnational in nature and operation. Pork is more and more a global industry with Chinese characteristics.

Why Is Pork the Heart of China's Meat Industry?

Pork is China's so-called national food (Wang and Watanabe 2008). More than the massive scale of production already described, this label reflects deeper cultural meanings and associations, and dietary norms and traditions. In Mandarin Chinese, the universal word for meat (*rou*) refers to pork. Formally, *zhurou* ("pig meat") means pork, but in conversation, on menus, and when ordering at a restaurant or market, talking about *rou* is talking about pork. By contrast, "chicken meat" (*jirou*), "cow meat" (*niurou*), and "sheep meat" (*yangrou*) must be specified: they are *types* of meat while pork *is* meat.

That pork stands in for meat is a current expression of a much longer history. Pigs and pork have been central to Chinese households for millennia, carrying cultural meanings and performing agroecological functions. For example, the Chinese character for home and family is 家 (*jia*). It was created some 3,500 years ago by adding the roof radical to the pig radical,[8] or more figuratively, by putting a roof over a pig's head (Harbaugh 1998). Similarly, there is a saying, *meiyou zhu, meiyou jia*, which translates as "no pig, no home." These examples illustrate that at some time, pigs lived in houses (or in the household) with families (Wieger 1927). They also reflect that pigs were thought of as key components of the household, and of the very meaning of "home."

In terms of diet, pork has been a consistent part of the major agricultural and food traditions in China for thousands of years,[9] but peripheral in diets until very recently. For most of history, and for most Chinese people, eating meat was limited to social and ceremonial events, never produced in quantities that would allow for routine consumption for the entire population. Before 1949, Chinese farmers received only 1 percent of their food energy from animal products, while grains made up the bulk of their diets (Hsu and Hsu 1977). While China today seems full of meat, pre-reform and pre-modern China was full of pigs. Starting 6,000 to 10,000 years ago, when pigs were domesticated in various parts of China, each place had its own locally adapted pig breed, and most households raised at least one or two pigs a year (Jian 2010; Zheng 1984). Pigs were more valuable alive than dead, acting as efficient converters of kitchen and agricultural scraps into nutrient-rich fertilizer (Schmalzer 2016; Wittwer et al. 1987), before becoming pork that could be given as a wedding gift, used to curry political or social favor, or eaten as part of Chinese New Year celebrations (Chang 1977). *Pigs* were a staple of Chinese farming systems and households, while for the vast majority of people *pork* was a rare treat. Although this long tradition of pork consumption in China included variation across different times, places, and social relations, the smallholder model of raising pigs as part of diverse crop and livestock agroecosystems, coupled with only occasional meat eating, defined much of the country's 7,000 years of agricultural history.

Meanings and associations have changed through time, but pigs and pork remain important cultural signifiers in China today. For instance, 2007 was a "Golden Pig" year in the Chinese zodiac. A symbol of prosperity and happiness, tens of thousands of couples in China were eager to have children born under this auspicious sign. Golden Pig fervor was so high that officials predicted a Golden Pig baby boom[10] and media outlets in China and internationally reported overloaded maternity care wards ahead of the New Year, as well as concerns about whether social institutions would be able to handle hundreds of thousands of Golden Pig babies thereafter (Cody 2007; *Economist* 2007).

Although other Chinese zodiac animals are considered auspicious, the 2007 Golden Pig was something special. Companies selling diapers, baby oil, and baby food increased their advertising budgets by 50 percent ahead of 2007 (Cody 2007), in hopes of cashing in on (and perhaps helping to create) the Golden Pig fertility rush. But more than short-term economic gains for baby companies, the frenzy around the 2007 Golden Pig year also suggests the importance of pigs in contemporary Chinese culture. The Golden Pig is a rich symbol of the new role that pigs play in China's economy. Pigs are big business: they occupy important positions in China's food and agricultural markets, they are a destination for state and private investment, they serve as a source of profit for agribusiness firms and the state, and concerted efforts to breed and feed them has rerouted trade, investment, and resource flows globally.

Why Is Pork Booming Now?

Although common and even ubiquitous in China for thousands of years, pigs in the reform era are dramatically different. From their earlier stature as symbols of home and suppliers of fertilizer, pigs have become most highly valued for the pork they produce. This is so in terms of economic value—with profits accruing to firms along the pork commodity chain—and in terms of social value—with meat consumption serving as a marker of social status. These transformations are part and parcel of China's broader capitalist transformations.

Throughout the reform era, meat in general and pork especially have signified progress against a backdrop of scarcity. During the Cultural Revolution (1967–1976) when pig production was collectivized, meat was rationed to households by coupon. Even though for centuries farmers only ate meat once or twice a year for holidays and special occasions, meat rationing impacted the diets of virtually all Chinese people, changing popular notions and expectations of the frequency and amount of meat consumption. What's more, the first thirty years of the People's Republic of China (from 1949) were punctuated with large-scale agricultural shortfalls, household-level food insecurity, and famine. During this time, many Chinese people directly experienced food scarcity.

Common experiences of rationing informed the state's reform-era focus on increasing pork production and consumption, both to heal the wounds of past scarcity, and to legitimize the state for its role in creating a bountiful agrifood system (Schneider 2017b). In a country where the politicians who run the central government are not elected, mechanisms for creating and sustaining state legitimacy are important, with productivist and environmental discourses becoming especially salient (Chen, Zinda, and Yeh 2017). Raising meat production is one such mechanism (with attendant discourses), making pork consumption a political goal, as well as a tool for economic development.

How Has the Pork Industry Developed?

Ramping up pork production since 1978 has entailed a suite of political economic transformations including liberalization, privatization, and commercialization broadly. In 1987, small-scale or "backyard" household farms supplied 95 percent of all the pork in China (Qiao et al. 2016). Today, medium-size and large-scale operations have the largest share, the result of years of concerted effort to scale up and modernize the sector, and to eliminate smallholding in favor of commercial enterprises.

Industrializing Pig Production

State support for industrializing the pork sector began early in the reform era, intensifying through the 1990s,[11] and solidifying as a goal and set of policies in the 2000s. Following earlier efforts to increase the scale and output of pork production,[12] industrialization got a boost from the PRRS (porcine blue-ear disease) epidemic that swept across China in 2006, leading to the cull of 20 million pigs and subsequent skyrocketing pork prices (An et al. 2011). In the wake of the outbreak, the central government doubled down on efforts to scale up and modernize pig production. The State Council laid out measures in 2007 to increase state support for large-scale, industrialized, and standardized pork production to stabilize the industry and protect against future shocks.[13]

Post-disaster subsidies successfully boosted the national swineherd from 40 million to nearly 50 million, igniting overall production growth. But subsidies not only increased production, they also restructured it. After dedicating 2.5 billion RMB ($366 million) to large-scale production facilities or specialized "pig barns" in 2009, the Ministry of Agriculture reported that farms raising more than fifty hogs a year accounted for almost 60 percent of total slaughter, an increase from less than half in 2007 (Woolsey and Zhang 2010).

Confined Animal Feeding Operations (CAFOs) "Pig barns" are various forms of confined animal feeding operations (CAFOs), ranging from relatively small-scale operations with 50–200fifty to two hundred pigs, to massive industrial megafarms raising thousands to hundreds of thousands of hogs each year. Regardless of scale, the logic of the CAFO is the same: confine livestock to control (and speed up) the conversion of feed to meat ("efficiency" in industry terms), while sparing land for other uses, freeing labor for other employment, and creating a sector that can be effectively monitored and regulated.

Discursively, CAFOs have been a rather easy sell. First, because China has 21 percent of the world's population and only 9 percent of arable land, CAFO production is seen as inevitable for raising meat consumption without compromising food security. Second, during the first decades of reform, economic development occurred primarily in cities on China's east coast, both pushing and pulling hundreds of millions of migrant workers from the farm to the factory (and the restaurant, construction site, massage parlor, KTV [Karaoke Tele Vision] lounge, etc.). Specialized pig production was proposed as a way to free many rural people from pig raising, while providing opportunities for some to concentrate on raising pigs exclusively (the medium-scale version), and for others to take up employment in commercial operations (large-scale production and processing). Finally, given rising concerns over environmental regulation and food safety, especially in recent years, CAFOs and large-scale operations are further justified as the most governable form of production. The discourse is that hundreds of millions of spatially dispersed peasants are difficult to regulate, while commercial farms are governable entities that can be more effectively monitored and controlled, especially through law (Schneider 2017b).

As in any place where CAFOs become the dominant form of production, claims about their efficiency and governability require concerted and critical analysis. Decades of research document the ecological crises of so-called factory farming, as well as social and public health implications (discussion follows). Research also demonstrates corporate consolidation around and through CAFOs, with firms effectively owning national livestock herds that live primarily in their factory farm operations (Howard 2016; see also chapter 2). While governments always play a role in corporate power through supportive policies and subsidy programs (Clapp and Fuchs 2009; McMichael 2009b), in China these relationships are not hidden; the state designates agribusiness firms as the leaders of agricultural development, and supports them as agents of economic and rural development. State-corporate relations are central to the political economy of China's reform-era pork boom.

The Political Economy of Industrialization

Industrialization in not a neutral process of simply capitalizing on economies of scale and responding to the changing labor composition in the countryside. It is also, and importantly, a process in which power relations structure the choices made about how "goods" and "bads" will be distributed both socially and spatially, creating a political economy of winners and losers. This political economy of the pork boom includes material relations among state, private, domestic, and international actors, who negotiate and manage pork industry developments and profits, and environmental relations on an uneven playing field. Today, the most powerful players in China's pork industry are domestic agribusiness firms, and government ministries and bodies at various political scales. These players are linked through industry groups, policy committees, and boards of directors. Together they are creating a robust domestic agribusiness sector to serve as an arena for national-level rural and economic development, and as a launching pad for accessing markets and resources abroad. Agribusiness-led vertical integration and industrialization are the hallmarks of China's modern agriculture (Schneider 2017a).

Modern Agriculture In the first decade of reform, agricultural productivity stagnated. Then in the mid-1990s, under the leadership of Jiang Zemin, the central government announced a development plan based on transitioning from traditional to modern agriculture. Government authorities characterized modern agriculture as commercialized (*shangyehua*), specialized (*zhuangyehua*), scaled up (*guimohua*), standardized (*biaozhunhua*), and internationalized (*guojihua*), and identified industrialization and vertical integration (*chanyehua*) as the primary means by which to achieve these goals. In 1998, authorities introduced a set of policies that put lead firms called dragon head enterprises[14] (DHEs) at the center of the modernization campaign. Central leadership began promoting and supporting these firms as *the* vehicles to bring about modernization (Zhang and Donaldson 2008, 29). Contracts between firms and farms became the preferred mechanism for integrating rural primary producers under the umbrella of modern agriculture.

Dragon Heads Agribusiness firms emerged rapidly after 1978, and subsequently, their number and importance have continued to grow. At the end of 2011, there were more than 110,000 officially designated dragon heads, and another 280,000 enterprises engaged in agricultural industrialization.[15] According to the State Council in 2012, "Dragon heads are not like ordinary commercial enterprises: they are responsible for opening up new markets, innovating in science and technology, driving farm households, and advancing regional economic development. They are capable of driving agricultural and village economic restructuring, driving commodity production development, promoting increased efficiency, and increasing farmers' income."[16]

The dragon head designation entitles a company to government programs that subsidize their rural and economic development responsibilities, and label the company as a lead firm, which can translate into enhanced legitimacy and trustworthiness in the market. State support comes in the form of direct subsidies for construction and operating expenses, as well as tax exemptions and reductions, export tax rebates, discounted loans with little or no interest, and investment (Zhang, Fan, and Qian 2005).

To become a dragon head, a firm must have legal standing as a state-owned or private enterprise, a group or corporation, a Chinese-foreign joint venture, or a wholly foreign owned enterprise.[17] It must also meet operational, financial, and farm integration criteria as outlined in government documents.[18] An important stipulation is that 70 percent of the firm's primary products processing and distribution must come through an "interest coupling mechanism" (*liyi lianjie fangshi*) for integrating farm households into their operations and markets. Officially, lead firms were working with 110 million rural households in 2011, using contracts with farmers, and following the "radiation-driven" (*fushe daidong*) farming model, in which technology, information, and market opportunities radiate from firms to cooperating or contracted farmers.[19] Unofficially, the figure is much lower, as studies find discrepancies between reporting and actualization of contracts between firms and smallholders.[20]

Restructuring the Pork Sector Government statistics report that dragon head-led vertical integration is the principal form of agricultural production, operating on 60 percent of the country's cropland, and covering 70 percent of livestock (pigs and poultry) and 80 percent of aquaculture production in 2011. Dragon heads' combined sales revenue was 5.7 trillion RMB (US$917 billion) in 2011, and they produced two-thirds of the average food basket in major cities (Guo, Jolly, and Zhu 2007).

The extent to which the dragon head model benefits small-scale farmers is unclear. It is certain, however, that smallholder numbers are declining. In 1996, 70 percent of agricultural households (135 million) raised pigs, with 77 percent of them raising five or fewer hogs each year. From 2002 to 2012, the number of farms raising fewer than 50 pigs per year decreased by half, while the number of farms with more than 50 pigs increased by 1.7 million. From 2009 to 2012 alone, the number of operations with more than 5,000 hogs grew from 8,300 to 11,400 (Gale 2017). In terms of marketed pork, farms with fewer than 50 pigs accounted 95 percent of slaughter in 1987, compared to only 35 percent in 2011 (Qiao et al. 2016). Reliable figures on the number of households raising fewer than 10 pigs (a "backyard" peasant farm) and their share of slaughter do not exist, though this scale is no doubt declining even faster than the undifferentiated "under 50" category.

Pork sector restructuring is the result of continued state support for the growth of corporate power and consolidation. Most recently, the Ministry of Agriculture's 13th Five-Year Plan for 2016–2020 set objectives for ongoing modernization. In the hog sector, objectives include further increasing the scale and vertical integration of hog farms, and shifting more control to companies (Gale 2017).

Feed Imports, Illusions of Land Sparing, and Offshoring

If 65 percent of China's 670 million pigs are produced on farms with 50 or more pigs, then more than 436 million Chinese pigs live in "pig barns," or CAFOs. These enclosed structures take up very little space. For an operation with 50 pigs, the building is about as big as the bodies of the 50 pigs themselves, with some wiggle room for feeding and tending the animals. Similarly, a 500-pig building is close to the size of 500 full-grown pigs, and a mega-operation with 100,000 pigs is made up of several buildings—and/or stacked "pig apartment" buildings—roughly the size of the number of animals they house. This is to say that the confinement system is just that: animals are confined together, tail-to-snout, with very little room to move. In large-scale operations, aside from pig stalls or crates, there is additional space for workers to carry out their tasks, for equipment, and perhaps for offices. The buildings, however, are small compared to their output, a relationship that is the hallmark of the CAFO's so-called efficiency.

Engineers and animal scientists have created a system that produces as much meat as possible from the smallest possible amount of space. This is accomplished in large part through breeding and feeding. Industrial pigs are made to survive the factory farm, with commercial feed mixes designed to make them quickly convert grains and oilseeds into muscle, even though they spend their lives standing in place. Modern breeding began early in the reform era, and now the same pig breeds that dominate industrial operations globally also dominate China's pork sector. "Exotic" pigs like Duroc, Landrace, and Yorkshires have all but replaced indigenous swine: exotics account for 90–95 percent of China's pork.[21] Modern feeding, too, began early in reform, when liberalization of soybeans co-produced the industrialization of pig farming.

Feed and Soybeans To increase pork consumption for upward of one billion people without converting the entire country into a massive pig feed farm, China's authorities had little option but to move toward industrial feeding systems. This began with efforts to develop feed milling in the early 1980s, first for pig feed, followed by chicken feed in the 1990s (Ministry of Agriculture of the People's Republic of China 2009). The real boom, however, came with soybean imports. Although China is the home of soybean domestication, and was the longtime world-leading soybean producer, liberalizing soy in the 1990s shifted not only domestic production and consumption, but also global production, markets, and flows.

Before Reform and Opening, soybean harvests in China were destined for tofu, soy sauce, and other human-food uses; they were rarely used for livestock feeding (Shurtleff, Huang, and Aoyagi 2014). Soybeans only became important for pork production in the context of China's globalization and capitalization. With the goal of raising meat consumption without sacrificing food security, and in anticipation of accession to the WTO in 2001 (which required selective liberalization), central authorities started to open soybean imports in the 1990s, while encouraging domestic processing. As a result, soy imports soared at an average growth rate of 26 percent. In 2016, China imported 91 million tons of soy, or 64 percent of the global trade, predominantly from Brazil and the United States. Imported beans, which accounted for 95 percent of soy crushing in China in 2016, are processed domestically to produce livestock feed (soybean meal) with soy oil as a co-product. These two agroindustrial uses now drive the country's soy industry, with global consequences: China is the world leader in soybean imports, in soybean meal production, and in soybean oil production (USDA FAS 2017b).

Offshoring Soy imports represent an offshoring of the needs and impacts of industrial pork production, as the land, labor, and resources that go into soybean production are largely managed in the United States and Brazil, each of which supplies around 40 percent of international soy trade (USDA FAS 2017b). The United States is the "traditional" world soybean leader, while Brazil is the new soy powerhouse, growing in concert with China's pigs and economy. Soybeans are Brazil's top export to China, balancing imports of Chinese consumer goods, and fueling new South America-East Asia coordinates in the global livestock-feed complex (Oliveira and Schneider 2016). Since 2009, China has been Brazil's largest trading partner, with soybeans accounting for 13 percent of all Brazilian exports in 2013 (Oliveira 2016).

It is important to note that these relations are not simply between nation-states. Indeed, as Gustavo Oliveira (2016) notes, the three largest transnational seed companies (Monsanto, DuPont, and Syngenta) control 55 percent of the global market for soy seed, and four trading companies (ADM, Bunge, Cargill and Luis Dreyfus, the first letter of each company name or surname forming the shorthand reference of "ABCD") control almost 80 percent of the international soybean trade. In South America, the latter four ABCD companies control 50 percent of crushing capacity and 85 percent of soybean exports. In China, foreign firms, including ABCD, own about 70 percent of soybean crushing (Yan, Chen, and Bun 2016), with domestic firms making up the balance.[22]

The lion's share of soybeans that feed China's pigs are the result of soil, water, and labor relations quite outside of China's borders. In a basic empirical sense, when the land needed to produce soybeans (and other feed crops) is included in the calculation of space needed for CAFO production, notions of the system's efficiency quickly evaporate (Schneider 2014; Weis 2010). Growing crops to feed industrial pigs requires billions of hectares of land and billions of gallons of water (Steinfeld et al. 2006). It requires the labor of people in all stages of feed production and transport, while creating surplus laborers by dispossessing people of their land and territory. It operates through corporate seeds and inputs (fertilizers and pesticides), supportive and coordinated national and trade policies, and often transnational feed industries. And it results in a host of social and environmental problems (e.g., Oliveira and Hecht 2016; Turzi 2011), which in this case, are also offshored to the global locations where farmers and firms produce the soybeans that fuel China's pork industry.

There are also, of course, social and environmental implications at home. As soybean imports have boomed, domestic soybean production has fallen drastically, sounding the alarm for a "soybean crisis" in the country. Between 2008 and 2013, the area of soy planted in China dropped by 24 percent, with the most dramatic reduction in Heilongjiang Province, which experienced a 42 percent drop (Yan, Chen, and Bun 2016). The livelihood implications of these changes are not yet well understood, but contribute to an overall hollowing out of the countryside and of rural social relations with the advance of capitalist industrial agriculture.

As Pork Booms, What Are the Social and Environmental Consequences?

Large-scale, intensive, corporate-led industrial meat production damages people, nonhuman animals, communities, and ecologies. The socio-environmental consequences of the global meat complex are well documented in policy and in academic, advocacy, and activist circles.[23] They are also increasingly well known among even casual observers who read mainstream international press such as the *Guardian* and the *New*

York Times.²⁴ From the climate-changing contributions of farting cows, mountains of manure, and transportation networks; to pathogens and antibiotic-resistant disease-causing organisms that make their way into waterways, food products, and human bodies; to the diversion of grains, land, water, and labor to feed livestock instead of people in a world full of food insecurity—the problems of industrial meat are no secret. And yet, CAFOs are the fastest-growing form of production in the world, steadily spreading through the global north and south, and the places in between. The growth of industrial livestock in China is not an exception, nor are its consequences; corporate concentration, public health crises, and environmental degradation are written into the model itself, wherever it touches down. The following sections briefly outline some of the most profound consequences of China's global pork industry, focusing inside of China.

Peasants and Corporate Concentration
Peasant production methods and relations are declining, while corporate ownership and concentration are increasing. Rather than distinct phenomena, the two are intimately linked: the definition of small-scale peasant production as "backward" is both a driver of industrialization and a result. The model of household pig raising that was practiced in China for thousands of years relied on locally occurring plants (including "weeds") and kitchen and agricultural "scraps" as feedstuffs. This is a slower growth cycle that produces a fattier pork, and in limited quantities; just enough for once- or twice-per-year pork consumption. In the rush to increase meat consumption, develop domestic agribusiness, and transition to a "modern" economy and society, the peasant mode of production and the peasant her/himself have come to stand in for what's wrong in China's food and agriculture systems, and what must be replaced. Denigration of the peasantry is certainly not unique to China or to the twenty-first century. But the specific register of this relation in China today is that political and popular discourse define the peasant as the problem for which further capitalist industrialization is the only and inevitable solution (Schneider 2015). Restructuring the pork sector—with large- and medium-scale operations replacing small-scale production—is an expression of these politics on the figure of the peasant.²⁵

In addition to scaling up and restructuring, corporate concentration is increasing. Dragon head enterprises (and agroindustrial firms more generally) control an ever-growing share of production, distribution, and retail. In the pork sector, three firms have led the industry in the past decade. In 2011, the annual sales of WH Group (formerly Shuanghui, or Shineway), Jinluo Meat Products, and Yurun Group accounted for 68 percent of total sales and 86 percent of total profits for the top ten pork processors in China. In addition to processing, these firms are vertically integrated from production through retail, and linked horizontally through industry and government organizations (Schneider 2017a). With WH Group's purchase of Smithfield in 2013, it is the world's largest pork processor, further adding to the concentration of corporate power in China's, and the world's, pork industries (see chapter 2). While the WH Group is the clearest example of Chinese pork firms operating as transnationals, it is not the only one. Jinluo, for instance, was incorporated in the British Virgin Islands, and conglomerated in Bermuda as a wholly owned foreign firm (Schneider 2017a). These examples reflect broader policy trajectories. Since 2000, the state has been encouraging firms to "go out" (*zou chuqu*) for access to markets and resources abroad, and to extend their global reach and competitiveness.

Dietary Change and Public Health
As agriculture changes, diets are also transformed. Chinese diets are becoming increasingly meaty, processed, eaten away from home, and associated with diet-related diseases. Again, meat consumption has quadrupled since 1980 to an average of 65 kg per person per year. Still, the increase is uneven across the population. According to official statistics, urban households bought an average of 36 kg of meat (pork, beef, mutton, and chicken) in 2012, while rural households consumed an average of 29 kg of meat, poultry, and processed products according to National Bureau of Statistics data. Unofficially, the difference is much greater. Because meals eaten away from home are not included in National Bureau of Statistics figures, middle- and upper-class urbanites who have disposable income to spend at an array of restaurants and urban eateries are consuming as much as three times more meat than rural residents (Xiao et al. 2015). Processed and packaged pork products are the fastest-growing market segments, sold increasingly in supermarkets and hypermarkets, as well as in family-owned shops and other small retail outlets. These products are also typically eaten away from home.

One result of dietary changes (of which industrial pork is an important part) is that public health is suffering. In 2015, cancer, heart disease, and cerebrovascular disease (hypertension) accounted for 69 percent of deaths in urban China and 68 percent in rural China. Compared to 1998 when the National Bureau of Statistics (n.d.) began reporting these data, the 2015 figures were an increase of 7 percent in urban areas and 20 percent in rural areas. At the same time, while China's pork has become leaner (because of commercial breeding and feeding), China's people are becoming fatter. A recent study in *The Lancet* found that there are more obese and overweight people in China than in any other place in the world. According to the study, more than 43 million Chinese men and 46 million Chinese women are obese, accounting for 16.3 percent and 12.4 percent of the respective global totals. Moreover, 23 percent of boys and 14 percent of girls under the age of twenty in China are overweight or obese (NCD Risk Factor Collaboration 2016). These "diseases of affluence" (Campbell and Campbell 2009) cannot be attributed entirely to diet (cancers especially are also related to pollution), nor can the rise of obesity be solely explained by increasing pork consumption. The pork production boom, however, is also a consumption boom, and an important component of changing eating habits, expanding waistlines, and the emergence of diet-related diseases and causes of death.

Environmental Health and Food Safety
The human health consequences of industrial livestock extend beyond diet. As elsewhere (Imhoff 2010), China is now experiencing the *environmental* health consequences of the CAFO. Food safety, which is the highest public concern in China (Song, Li, and Zhang 2014) and a key area of focus for the state,[26] is the clearest illustration. One example is growth promoter, or "lean meat powder" (*shouroujing*) residue in meat. The most highly publicized case was in 2011 when pork from Shuanghui (now WH Group) was found to contain clenbuterol, a banned pork industry growth promoter that can cause toxicity-related illnesses and cancers in humans. Another example is the emergence of antibiotic-resistant disease-causing organisms in meat products, soils, and water. Because CAFO operators administer antibiotics in "subtherapeutic" doses throughout the production cycle to promote growth, antibiotic-resistant strains of bacteria develop, making their way into manure, the environment, and human bodies. This is a global problem

that is particularly acute in China where nearly half of the 210,000 tons of annually produced antibiotics end up in livestock feed (Hvistendahl 2012), and 25 to 75 percent of those antibiotics are "excreted unaltered in feces and persist in soil after land application" (Luo et al. 2010, 7220). Antibiotic resistance—especially to tetracycline, which is most used in CAFOs—is a serious and growing public health problem in China. Antibiotic-resistant genes have been found in soils around feedlots, in water, and in human guts (Hvistendahl 2012; Ji et al. 2012; Luo et al. 2010). They are accompanied and complemented by heavy metals, especially mercury, copper, and zinc (Ji et al. 2012).

CAFOs also impact ecosystem health. Since 2010 when the government released results of China's first national pollution census, it has been widely known that manure from industrial livestock is the country's biggest source of water pollution.[27] Industrial pig and chicken operations dump manure onto soil and into waterways, such that phosphorus and nitrogen levels exceed what can be cycled and recycled. Waterways become eutrophic, killing some species of aquatic life while promoting others, and reducing the amount of water available for rural households. Manure in water, combined with fertilizer runoff from crop fields, is so severe that a dead zone has developed in the East China Sea, at the convergence of the *Changjiang* (Yangtze) and *Huanghe* (Yellow) Rivers (Diaz and Rosenberg 2008).

Food safety, public health, and water pollution are the pig-related environmental issues highest on the Chinese government's agenda. They are not, however, the only problems of the CAFO. Species diversity is declining as industrial hogs replace indigenous pigs. Greenhouse gas (GHG) emissions from manure and transport grow in concert with the industry (see chapter 7). Particulate pollutants fill the air around pig barns. And perhaps most infamously, dead pigs float in rivers.[28] How these issues are resolved—or not—will have further consequences for the trajectory of the pork industry, and China's people, land, water, ecology, and politics.

Conclusion: What's Next?

China's pork boom is a political economic process that capitalizes on (and transforms) cultural meanings and historical practices, while rerouting global flows of capital, commodities, and harm. It is a boom of world-scale proportions with global and local consequences. It is a boom that has increased meat consumption for hundreds of millions of people, and profits for a generous handful of Chinese and foreign firms. It is a boom with political import, both legitimizing the party-state and economic development, and drawing harsh public criticism over its public health and environmental degradations. And it is a boom almost forty years in the making, with an uncertain ending date.

By way of conclusion, two recent developments are important for considering the trajectory of China's global pork industry. First, pork production has stagnated in China since 2015, equating to a corresponding global production decline. As a result, imports have increased, much to the delight of exporting countries and firms. In 2016, in addition to producing and utilizing 45 percent of the world's pork, China also imported 31 percent of total pork imports (USDA FAS 2017a). Some market analysts point to stricter enforcement of environmental regulations in China as the cause behind the pork downturn, as operations in urban areas especially have been forced to close or relocate (FAO 2017b). Others cite price (labor scarcity and rising feed and pork prices) in China as the major force driving imports and creating market opportunities for exporters (Gale 2017). In either case, analysts predict further expansion of the pork sector though ongoing restructuring with "enhanced efficiency and greater economies of scale" (FAO 2017b, 45). In other words, more industrialization, more vertical integration, more and bigger CAFOs, more agribusiness control, and fewer smallholder farmers.

Second, and somewhat alternatively, there is growing recognition that (industrial) pork cannot and should not boom forever, given public health and environmental consequences. For instance, in 2016, China's Ministry of Health released new dietary guidelines calling for a 50 percent reduction in meat consumption.[29] Although guidelines are not binding policies, the new language around "reduction" indicates at least discursive acknowledgement of the health and climate problems that have accompanied pork's rise. At the same time, Chinese consumers' food safety concerns are compounded by increasing awareness of the health and environmental impacts of meat consumption. Some studies suggest that middle- and upper-class urban consumers have more faith in imported and/or industrially produced meat, which they view as more regulated and therefore safer than pork produced domestically and by peasants (De Barcellos et al. 2012; Xiu et al. 2017). However, not all consumers who have incomes high enough to make consumption decisions share this point of view. Alternative markets have emerged that offer ostensibly healthier and more "sustainable" meat and food products. Among the blossoming "alternative food networks" in China—which include farmers' markets, community-supported agriculture (CSA), and organic and biodynamic farming—consumers also cite food safety as their primary motivation for participation (e.g., Shi et al. 2011). These markets are often more locally based, using their nonindustrial nature as a selling point and a tacit point of critique.

Whether government guidelines, consumer choice, and alternative markets necessarily lead to less industrial meat is an open question. That there are distinct class politics involved in meat and meat markets is increasingly clear. Perhaps industrial pork will continue to grow in the coming years, penetrating even further into the layers of Chinese society, reaching more rural and low-income people, and further restructuring their diets. Perhaps organic, local, biodynamic, and "boutique" pork will develop further for especially middle- and upper-class people and markets. And perhaps some smallholder farmers will continue to raise pigs under more-than-challenging social, economic, and environmental conditions, maintaining pigs' genetic diversity and occasionally eating fresh pork in addition to processed sausages. In any case, China's global pork industry is firmly rooted for the time being, even if its foundation is as shaky as it is world shaking.

Notes

1. FAO statistics on global meat production begin in 1961.

2. See figure 1.2.

3. Unless otherwise noted, production and trade statistics are summarized from FAO 2019.

4. According to the most recent estimates from the USDA FAS (2017b), China produced 48 percent of the world's pork in 2016, reflecting a decrease from the most recent peak in 2014 when China produced 51 percent of global pork.

5. While average annual consumption in places like the United States (120 kg) and Australia (118 kg) dwarfs the Chinese figure, China's meat consumption is well above the world average of 42 kg of meat per person per year, and is expected to continue to grow (Weis 2013a).

6. Meat consumption was calculated by adding FAO meat production and import figures and subtracting export figures. Per capita consumption for 2017 was calculated using population figures from China's National Bureau of Statistics (n.d.).

7. Material in the chapter draws from eighteen months of fieldwork on pork industry transformations in northeast, southwest, and southern China from 2009 to 2012. Ethnographic material is supplemented with secondary data from media outlets, scholarly journals, organizational documents, agribusiness and government websites, and government and organizational statistics.

8. Radicals are simple characters used as building blocks for making more complex characters (Wong 1990). 家(*jia*) is the combination of the roof radical and a condensed version of the character for pig 猪(*zhu*).

9. For detailed treatments of the uses and significance of pork in Chinese cooking, eating, and cultural organization, see Chang (1977) and Anderson (1988). Also note that Chinese Muslims—including the Hui, Uyghur, Kazakh, Dongxiang, Kyrgyz, Salar, Tajik, Uzbeks, Bonan, and Tatar minority groups—do not eat pork in any form.

10. Reports differ on whether there was, in fact, a Golden Pig baby boom. Data for 2016 from the National Bureau of Statistics (n.d.) suggests there was not, while some media reported Golden Pig births were double that of a normal year (Li 2010). There is also disagreement about the extent to which an uptick in births could be attributed to the Golden Pig, as opposed to a demographic bump related to the mini-baby boom in the 1980s (People's Daily 2007).

11. For a discussion of agricultural policy in the 1990s, see Zhang and Donaldson 2008.

12. See Gale 2017, 6, for a summary of recent pig-related policies.

13. Measures included: direct farm payments for sow insurance and disease prevention, compensation for losses from the PRRS epidemic, subsidies for seed-breeding of live pigs, investments in production infrastructure and market-monitoring systems, grants for safe disposal of sick pigs, rewards for counties that significantly increased production, and financial incentives for leading agribusiness firms.

14. Throughout the chapter, I use *dragon head, dragon head enterprise, lead firm,* and *leading firm* or *enterprise* interchangeably. All signify the word *longtou qiye* in Mandarin Chinese.

15. These figures are from the inaugural speech of Hui Liangyu, deputy prime minister of the State Council, at the launch of the China Association of Leading Enterprises for Agricultural Industrialization in 2012. The full text of the speech (in Chinese) is available at http://baike.baidu.com/view/9676144.htm.

16. Author's translation of the Baidu Baike entry for "dragon head enterprise" at http://baike.baidu.com/view/125729.htm (in Chinese).

17. Although possible, foreign firms are not typically designated as dragon heads. See Schneider 2017a.

18. Eight government institutions jointly issued the "Provisional Measures for the Administration of Dragon Head Enterprise Identification and Operation Monitoring" in 2003, with an update in 2010. The full text (in Chinese) is available at http://wenku.baidu.com/view/090cc1e96294dd88d0d26be4.html.

19. From Hui Liangyu's 2012 speech (note 15).

20. See Guo, Jolly, and Zhu 2007. I also found evidence of firms over-reporting their relationships with smallholders during my fieldwork.

21. Interview No. 58, Ministry of Agriculture, Beijing, September 17, 2010. The native pigs that remain are raised either by small-scale farmers, on specialty "boutique pig" farms, or on state-funded and largely privately run conservation farms tasked with preserving genetic diversity.

22. See Oliveira and Schneider (2016) for a discussion of how transnational corporations become leaders in China's soy sector after the soybean-crusher enterprise defaults in 2004. See Schneider 2017a for an analysis of dragon heads and state investment in domestic soybean firms. And see Yan, Chan, and Bun 2016 for a discussion of the crisis of domestic soybean production and farmers in the context of soaring imports.

23. Leading policy publications include *Livestock's Long Shadow: Environmental Issues and Options* (Steinfeld et al. 2006) and the Pew Commission's report *Putting Meat on the Table: Industrial Farm Animal Production in America* (2008). Recent academic work includes *The Ecological Hoofprint: The Global Burden of Industrial Livestock* (Weis 2013a); *Every Twelve Seconds: Industrialized Slaughter and the Politics of Sight* (Pachirat 2013); *Political Ecologies of Meat* (Emel and Neo 2015); and several articles in academic journals. Among the many advocacy and activist books and references are *Meat Atlas: Facts and Figures about the Animals We Eat* (2014), published by the Heinrich Böll Foundation and the Friends of the Earth Europe, and *The Meat Racket: The Secret Takeover of America's Food Business* (Leonard 2014).

24. The *Guardian* has published several pieces related to the problems of industrial livestock production; see, for example, Harari 2015 and Vidal 2010. Similarly, see Bittman 2008 as well as the *New York Times* "Factory Farming" news index page at https://www.nytimes.com/topic/subject/factory-farming.

25. For thorough analysis on the figure of the peasant in contemporary China, see Day 2013.

26. Food safety was on the state's agenda in the 12th Five-Year Plan (2011–2015), and is again prominent in the current 13th Five-Year Plan (2016–2020).

27. See China Pollution Source Census 2010 (in Chinese).

28. The most serious instance was in March 2013, when 16,000 pig carcasses were found floating in the Huangpu River in Shanghai. Other similar, though less extensive incidents have occurred since. See Duggan 2014.

29. China's new dietary guidelines can be found at http://dg.en.cnsoc.org/ (in English and Chinese).

5 Amerindians, Mestizos, and Cows in the Ecuadorian Amazon: The Silvopastoral Ecology of Small-Scale, Sustainable Cattle Ranching

Thomas K. Rudel

Denny (see chapter 7) ends her global survey of greenhouse gas (GHG) emissions from livestock with the hope that cattle ranching, with particularly high rates of emissions, might be reformed in such a way as to reduce its emissions intensity. The following pages describe a silvopastoral path to reforming cattle ranching that would reduce its emissions intensity. A case study of a spontaneous conversion to a silvopastoral landscape in the Ecuadorian Amazon illustrates the promise of this path to lower-emissions cattle ranching. The analysis begins with a brief exploration of the socio-ecological conditions that foster resource conservation.

Economists have long maintained that high discount rates encourage resource users to engage in ruinous exploitation of natural resources (Clark 1973). If the value of a resource ten years from now is only a small fraction of its value to an investor today, she will probably be looking to exploit that resource as much as possible in the short-term, then abandon it, and move on to exploit another resource with more lucrative prospective returns to an investment. Applied to a natural resource like land, high discount landholders anticipate that they will stop using a tract of land in a few years, so they will do little to conserve the productive potential of their lands for future growing seasons. Alternatively, a landholder with low discount rates will value the productivity of the land ten years from now almost as much as the value of the land this year. She does so because she anticipates earning her livelihood from these lands in ten years. For this reason, the low-discount-rate landholders will implement conservation measures designed to maintain the productivity of the land.

These calculi translate directly into different postures towards resource conservation among landowners and land renters. The renters, who may not cultivate a rented tract of land in the coming growing season, have few incentives to engage in conservation practices. Landowners, with longer time horizons, may value their land in ten years nearly as much as they value it right now, in part because landowners, confronted with the inevitable uncertainties of selling their land, plan to reside on it and earn their livelihood from it for an extended period of time. Studies from a range of settings, the American Midwest (Carolan 2005), the Amazon basin (Rudel, Katan, and Horowitz 2013), and a worldwide meta-analysis of tropical deforestation (Robinson, Holland, and Naughton-Treves 2014) all confirm that individuals who own their land are more likely to embrace environmental stewardship than people who rent land, use government land, or use open-access lands.

More broadly, this line of reasoning about the affinity between landownership and environmental stewardship of the land is consistent with arguments about the social bases for sustainable agriculture. Netting (1993) in his work on the ecology of small-scale, sustainable agriculture assumes that smallholders own at least a large portion of the land that they use, that they work full time on their farms, and that they think in terms of long-term occupancy of the land. In this world, small-scale ranchers would make environmental improvements through laborious inputs of labor like the digging of drainage ditches, the planting of trees, and the recycling of manure. Landowners can carry out these tasks because they work full time in agriculture. They also expect that either they or their families will continue to work these lands for the foreseeable future, so they or their descendants will live to see the benefits of their labor in sustaining production on their lands.

In sum a positive association between owner-occupied pasture lands and the environmental stewardship of these lands seems both theoretically plausible and empirically supported in at least some instances. The difficult question concerns countervailing trends. Two of them, more globalized markets and more part-time farming, are prevalent throughout Latin America. Their prevalence could curtail the spread of sustainable agricultural practices in this region. The socio-logic behind these countervailing trends is outlined as follows.

Globalization has characterized cattle ranching in Latin America over the past thirty years. Markets have extended across regional and international boundaries. Shipping of Brazilian beef to Europe began after the

Brazilians eradicated hoof and mouth disease at the end of the last century (Nepstad, Stickler, and Almeida 2006). Argentine beef began to be sold in large quantities outside of the Southern Cone of South America, especially when the economic collapse of 1998–2002 led to the rapid devaluation of the Argentine peso against the dollar and made Argentine exports like beef very inexpensive outside of the country. With the arrival of cheap Argentine beef on the shelves of supermarkets, ranchers in an importing country like Ecuador have few ways to preserve market share. An increase in the scale of operations offers one potential way to survive in a newly globalized market because inputs like veterinarian services cost less per animal. Ecuadorian farmers opted to take this path: renting pastures from other landowners offered a relatively inexpensive means of expanding the scale of cattle operations, so the extent of rented pastures increased as the pressures of globalization became more palpable. In this way globalization would indirectly reduce the incentives for environmental stewardship among cattle ranchers.

The logic that ties landowning cattle ranchers and their livestock to sustainable practices also begins to break down when farming becomes a part-time occupation and farm families decline in size. The work unrelated to farming reduces working hours on the farm. It may also encourage off-farm residence that in turn could reduce the time available for farm work if getting to the farm entails a commute. Renting pastures to others may become appealing in this context. With farm labor scarce, the attractions of sustainable regimens like agroforestry or silvopastoral ranching diminish because farmers cannot find the labor necessary to plant the trees and the improved forages that are integral to sustainable practices (Calle, Montagnini, and Zuluaga 2009; Dagang and Nair 2003; Rao et al. 2015). For these reasons growth in part-time farming could be associated with a reluctance to adopt labor-intensive sustainable practices on farms or ranches.

How do we sort out the magnitude of these opposing, intersecting trends of globalization, part-time farming, and stewardship norms spreading among small-scale cattle ranchers? A case study of the influence of rental versus owner-occupied pasture management in a context marked by globalization and the spread of part-time farming might enable us to tease out the relative magnitude of these opposed influences and, in this light, assess the prospects for sustainable cattle ranching in Latin America. The case under study involves small-scale cattle ranchers in the Ecuadorian Amazon. Some of these smallholders are *mestizo* migrants from the Andes whose families settled in the Amazon region forty to fifty years ago. Others are *Shuar*, lowland Amerindians, who, faced with encroachment and invasion of the Amazon lowlands by Andean migrants, converted from shifting cultivation to cattle ranching in the twentieth century in order to acquire secure titles to land. In the concluding section, after reporting on the opposed, intersecting trends toward more globalization, more part-time farming, and more stewardship on pasture management over a 30-year period among Shuar and mestizo cattle ranchers, I address the generality of these social and ecological dynamics across Latin America as a whole.

The Context: Small-Scale Cattle Ranching in the Ecuadorian Amazon

The people under study reside near Macas, in the upper reaches of the Amazon basin in Ecuador, just to the east of the front range of the Andes at about 1,000 meters of elevation (see figure 5.1). Tropical rainforest covers the unexploited portions of the landscape. It rains a lot! So much so that slash and burn methods of shifting cultivation do not work in this area because the forests are too wet to burn. "Slash and rot" rather than "slash and burn" described the pre-contact agricultural economy of the Shuar. Rivers run eastward into tributaries of the Amazon. The soils are acidic and infertile. The only people living in the region prior to 1950 were lowland Amerindians (the Shuar), a small number of Catholic missionaries who had labored to convert the Shuar to Christianity, and a few mestizos with homes around the missions. The Shuar lived in isolated homesteads in the rain forest. A small stream of mestizo migrants from the Andean highlands had begun to move downhill into the Amazon during the first half of the twentieth century. At first, the mestizos settled near the missions, but, as their numbers grew, they moved outward and displaced Shuar families from their ancestral lands.

To prevent the dispossession of the Shuar from their ancestral lands, the missionaries began to advocate for a change in Shuar settlement patterns. Shuar would resettle in villages and file, as communities (*centros*), for a collective title to the 4,000 to 5,000 hectare tracts of land around each of the newly created villages. The Shuar became quite adept during the 1970s and 1980s in creating new settlements and lodging claims to the surrounding land. Mestizo landholdings also expanded throughout the 1970s and

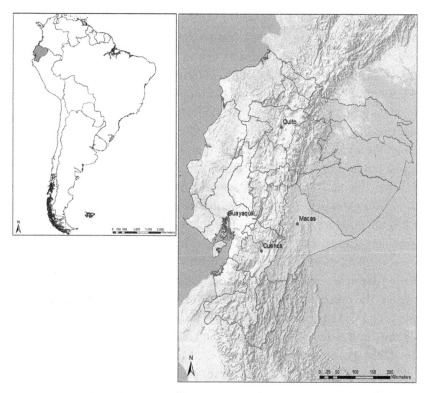

Figure 5.1
Ecuador study area: Morona Santiago

1980s. Mestizos too lodged claims to land, but as individuals. To strengthen their claims to forested land during the 1960s, 1970s, and 1980s, the Shuar followed the land-clearing practices of mestizos. The Shuar cleared patches of claimed land, planted pasture grasses, and acquired small herds of cattle (Rudel and Horowitz 1993).

By the end of the 1980s, the mestizos and Shuar had occupied almost all of the arable land in Morona Santiago. The Shuar had title to 42 percent of this land, and mestizos had title to the remaining 58 percent. While groups of Shuar had collective titles to blocks of land that contained village centers, individual Shuar households within each group took possession of particular tracts of land within their block of land. These tracts of land could be passed on to heirs, so in several vital respects Shuar households had acquired recognizable, individual agricultural smallholdings of 50 to 70 hectares by the 1980s.

In other respects, the position of the Shuar remained distinct from the mestizos, particularly in their access to credit. Because the Shuar did not have individual titles to land and because the *centros* would not allow individual Shuar to sell their landholdings to non-Shuar, the Shuar could not use their land as collateral to obtain loans from banks, so Shuar cattle ranchers found it impossible to take out loans to purchase cattle. In sum, by the late 1980s, both mestizos and Shuar had established small-scale cattle ranches throughout the valleys at the eastern base of the Andes, but their access to funds to finance cattle ranching remained unequal.

Over the past two decades, modest amounts of urbanization have occurred in Morona Santiago, especially around Macas, the provincial capital of the province. Growing employment in non-farm-related work in the provincial capital of Macas also created incentives for landowners with non-farm jobs to rent their lands to other cattle ranchers, so the extent of rented pastures would have been expected to grow with urbanization in the region. The province, about the size of the state of Vermont, had a population of 115,413 persons in 2010. The urban center of Macas had 19,176 persons in 2010 while the entire municipality, including extensive rural areas, had 41,155 persons (Censo del Ecuador 2011).

Local and provincial governments had no powers of taxation and remained dependent on the central government for funds to operate basic services like schools. After 2005, the central government focused its expenditures on upgrading the roads in the region. By 2015, newly paved roads connected most parishes with town centers in the valleys. Some people had begun to manufacture finished goods like furniture using wood from the forests of the region. Fertility rates had dropped significantly after 1980, first among mestizos and ten years later among the Shuar. Infestations of pests during the early 1990s reduced the size of the *naranjilla* crop (a citrus fruit) and, in so doing, reduced rural incomes substantially. In response, large numbers of young mestizos, but not young Shuar, left the region to look for work in urban areas or overseas. In most cases the emigrants left any lands they owned in the hands of their relatives.

The macroeconomic context in Ecuador remained volatile during the late twentieth century. In 1998, in the midst of an economic crisis, the country abandoned its own currency and adopted the U.S. dollar as the currency of choice for the Ecuadorian economy. An expansion in imported goods followed shortly thereafter as the stronger and more stable dollar made goods like Argentine beef, produced in countries with currencies that were depreciating against the dollar, very inexpensive in Ecuador. To curry favor with urban consumers, beleaguered politicians in Ecuador's central government eliminated tariffs on imports of Argentine beef. The sudden access of low-cost Argentine beef to the Ecuador market created an economic crisis for Ecuadorian cattle ranchers. Given the low prevailing prices in markets for beef, ranchers in the Ecuadorian Amazon lost money on each cow that they raised and took to market. This circumstance led many small-scale ranchers to liquidate their herds in the early 2000s. Since then the prices for beef have recovered, and Ecuadorian landholders have tried to rebuild their herds. The economic and environmental consequences of these contextual changes are partially visible in the patterns of change captured in three surveys of cattle ranchers in Morona-Santiago between 1986 and 2011. I will outline these changes and explore their implications for land ownership and environmental stewardship in Amazonian pastures.

During the past twenty years, the cattle pastures in some portions of Morona Santiago appear to have changed in a fundamental way. The initial pattern of land clearing in the region during the mid-twentieth century involved clear cutting. Old photographs of these pastures show few trees, if any. Smallholders did not leave trees standing in fields that they sowed with pasture grasses for fear that the trees would fall on cattle and kill them during the violent thunderstorms that occur in the region. These treeless pastures have begun to give way selectively to silvopastures populated by large numbers of small trees in addition to the pasture grasses. By 2011 the numbers of trees in each hectare of pasture ranged from an average of 85 in one community to an average of 358 in another, nearby community. The communities with the highest densities of trees in pastures also had the largest proportion of small, recently germinated tree stems. This pattern suggests that the growth in the density of trees in pastures has occurred recently. By extension, the volume of GHG emissions associated with cattle ranching has begun to decline as the carbon sequestration of the trees in pastures has increased. What, then, has driven this trend towards a more silvopastoral landscape?

Three surveys of small-scale cattle ranchers conducted over a twenty-five-year period make it possible to answer this question, at least in part. Each survey contains subsamples of Shuar and mestizo cattle ranchers. The ethnic composition of the respondents in each survey varied from approximately 33 percent Shuar in 1997 and 2011 to 50 percent Shuar in 1986. All of the data were collected in face-to-face interviews with respondents. The respondents were all asked the same questions, but in each survey the interviewing of people in different subsamples was done by different persons. Tensions between Shuar and mestizos made it important to use friends of Shuar or Shuar themselves in carrying out the interviews in Shuar villages. In effect, we carried out three repeated cross-sectional surveys of small-scale cattle ranchers in Morona-Santiago. By comparing the cross-sectional differences between the respondents in 1986 with those from 1997 and 2011, it becomes possible to trace out the socio-ecological changes occurring in this cattle ranching region of the Ecuadorian Amazon (Firebaugh 2008).

The Patterns of Change: Globalization, Urbanization, and Spontaneously Occurring Silvopastures

Continuity as well as change has marked the cattle economy of Morona Santiago over the past three decades. Of course households in the farming communities of Morona Santiago had other potential sources of income besides cattle ranching, especially in places like those surrounding Macas where some urbanization has occurred over the past thirty years. Perhaps because of the relative remoteness of the region from major centers of population in Ecuador and the corresponding absence of roads, cattle ranching has persisted as the most common source of income for a substantial number of households. Cows are a particularly attractive option for landowners with no direct access to roads because cows, unlike a crop, will walk out to a road before they are sold to a middleman and shipped to a market. In the interviews 58.7 percent of all households in 1986, 37.5 percent of all households in 1997, and 47.4 percent of all households in 2011 reported the sale of cattle as their chief source of income. In sum while the Shuar devoted more of their land than mestizos to cultivating root crops like taro and fruits like narangilla for sale in urban markets (2.1 versus 1.5 hectares, $p < 0.10$), cattle ranching has remained a mainstay in the household economies of the region throughout the post-settlement period.

The changes in landholdings and cattle herds in the region over the course of twenty-five years are outlined in table 5.1. Several trends are readily apparent. First, the mean size of the landholdings declined from 60 to

Table 5.1

Trends in landholdings and cattle herds, 1986–2011: Three surveys

		1986	1997	2011
Landholdings (hectares)	Mean size	60.5	50.7	30.7
	Kurtosis	2.1	2.9	28.3
Head of cattle	Mean size	14.3	11.0	17.9
	Kurtosis	0.9	1.1	7.7

Source: Interviews with rural household heads: 1986, 1997, and 2011.

30 hectares while the herds of cattle remained approximately the same. Stocking rates of cattle in pastures went up. The subdivision of lands that occurs with the retirement or death of the first generation of landowners accounts for most of the observed changes. The decline in the size of landholdings was more pronounced among the Shuar than among mestizos. Over the twenty-five-year period the mean size of mestizo landholdings declined from 57 to 42 hectares while the mean size of Shuar landholdings declined from 66 to 22 hectares.

Differences in migration patterns among young mestizos and young Shuar may account for these different trajectories of change. Many of the younger mestizos migrated elsewhere to work in the 1990s, and they had little interest in inheriting land, so often there was only one heir to land, and in that circumstance the land passed from one generation to the next without being subdivided. Fewer of the younger Shuar migrated elsewhere to work in the 1990s. Staying closer to the homestead, they were more interested in receiving their "fair share" of land at the time of inheritance. The persistence of higher fertility among the Shuar into the 1990s also meant that there were more heirs in Shuar households than in mestizo households, so at the time of inheritance Shuar lands often had to be divided in more ways than did mestizo lands.

The other dramatic change in cattle and landholdings occurs after 1997 and is signaled by the dramatic increases in kurtosis in both the distribution of land and in the distribution of cattle between 1997 and 2011. Kurtosis measures the length of the tail in any distribution of values, so the increase in its size after 1997 signals an increase in the inequality of landholdings and cattle herds. A small number of ranchers began to work more than two hundred hectares of land and maintained herds of more than one hundred cattle. During the initial period of settlement, restrictions on the size of landholdings set by the Ecuadorian state's agrarian reform and new land settlement agency, *el Instituto Ecuatoriano de Reforma Agraria y Colonizacion* (IERAC), limited the variance in the size of landholdings both among and between mestizo and Shuar landowners. This restraint on the accumulation of land and cattle disappeared in 1994 when the agrarian reform agency was absorbed into an institute for agricultural development and for all practical purposes ceased to operate (Immigration and Refugee Board of Canada 1999).

The change in currency in the Ecuadorian economy in 1998 disturbed the cattle economy in Morona Santiago. To curb rampant inflation in the Ecuadorian currency, caused in part by a slump in oil prices during the late 1990s, the government decided to abandon the *sucre* and adopt the dollar as Ecuador's currency. As noted, this shift from a weak (the sucre) to a strong (the dollar) currency altered the goods that Ecuador could import. In particular, the adoption of a stronger currency in Ecuador at the same time that Argentina's government leaders abandoned their attempt to peg the value of the Argentine peso to the dollar made it possible to import previously unaffordable Argentine beef into Ecuador at a price that significantly undercut the price of Ecuadorian beef. In this context the Argentine beef quickly captured a major share of the urban market for beef in Ecuador. In this circumstance the prices for Ecuadorian beef in local markets declined to the point, as noted earlier, where Ecuadorian cattle ranchers lost money on each cow that they sold. To prevent these losses, Ecuadorian producers, both Shuar and mestizo, liquidated their herds. In effect they "destocked" their lands and turned to other pursuits to make money. This cross-border dynamic drove small Ecuadorian ranchers out of business, which further opened up market opportunities for larger Argentine ranchers to export their beef. It resembles on an international scale the agglomeration that occurred in the Chinese pork industry during the last two decades of the twentieth century (see chapter 4).

Several years later, when it again became profitable to raise cattle and sell them locally, landowners began to raise money to reestablish their herds. Of course access to capital to restock pastures varied, sometimes dramatically, from landowner to landowner. Those landowners with sufficient capital sought to distinguish themselves from neighboring ranchers by emphasizing the higher quality breeds of their cattle. By 2011 some growers had taken to importing bulls from the United States to improve the quality of their herds. Others boasted that their cattle had caught the attention of cattle breeders from more economically established areas like the coast of Ecuador. In sum, landowners were not only restocking their pastures after 2000, they were also seeking to distinguish their cattle from those of their neighbors with the expectation that their brand of cattle would attract higher prices in a more globalized market for beef. These trends in restocking pastures accentuated the growing inequality in the size of farms and herds that is visible in table 5.1.

Mestizos had decided advantages over the Shuar in the restocking of pastures in Morona Santiago after 2000. Because larger numbers of young mestizos had gone overseas to work, they remitted more money to their elders, and these funds could be used to purchase more cattle. The other crucial difference between mestizo and Shuar landowners involved access to credit from banks. Mestizos mostly had fee simple "pre-titles" to their land that enabled them to get state-subsidized loans from the agricultural development bank (*Banco de Fomento*) in which they put their land up as collateral to secure the loan.[1] If they fell behind on the loan repayments, the bank could foreclose on the loan and take possession of the land. Individual Shuar, with only collective titles held by entire village, could not use their *centro*-designated tract of land as collateral for a bank loan. Furthermore, the legal prohibition on the sale of Shuar lands to non-Shuar persons or entities meant that state banks could not take possession of a Shuar borrower's lands in the event of a foreclosure. Aware of the deleterious effects that restrictions on credit imposed on economic development (De Soto 1989), the Salesian (Catholic) missionaries had created a revolving fund for bank loans for the Shuar in the 1970s, but the amount of money available to Shuar borrowers through this source remained small in the late twentieth century. These differences in access to capital made it possible for mestizos, but not for Shuar, to rebuild their herds of cattle fairly quickly after 2000. By 2011 mestizo herds averaged 20.4 head of cattle while Shuar herds averaged only 2.8. With land but no cattle, the Shuar had incentives to rent their pastures to mestizos with cattle and earn at least a small sum of money (in 2013 around $15 per month per hectare of pasture).

The changing incidence of rented land among Shuar and mestizos in table 5.2 testifies to the diverging trajectories of these two groups of small-scale cattle ranchers and suggests economic and environmental differentiation between mestizos and Shuar. The relative ease with which people could acquire land during the immediate post-settlement period of the 1980s had depressed the market for rented land. Why rent land when you could, with some effort, acquire your own land by occupying unclaimed lands several hours' walk from a road? For these reasons the demand for rented lands during the 1980s was low, and few landowners among either Shuar or mestizos rented out their pastures to others.

By 2011 a new pattern of land rentals had emerged (see table 5.2). More people were renting their pastures to others, particularly among Shuar landholders. The ethnic differences in the incidence of renting are statistically significant. Almost three-fourths of all Shuar landholders were renting their pastures to other people with herds of cattle and insufficient pasture to maintain them. These differing incidences of rented land had important environmental consequences for the two groups of smallholders.

With primary and secondary forests adjacent to many of the Amazonian pastures, seed rain from the nearby patches of forest tends to be heavy, so seedlings sprout spontaneously in the pastures. Owners or renters first encounter the seedlings when they "clean" the pastures after the cattle have eaten the forage. The cleaning consists of cutting down any weeds, brush, or low-value trees that have sprouted in the pasture since the last cleaning, which may have occurred a year earlier, right after the cows last grazed an area. The cleaning enhances the productivity of pasture grasses in the next season by eliminating other plants that would compete with the pasture grasses for sunlight, water, and soil nutrients.

At least in theory, renters and owners might adopt somewhat different stances toward the cleaning of pastures. Renters primarily will be concerned

Table 5.2

The changing ethnic profiles of people who rent out pastures to others

	1986	2011
Mestizo	16%	35%
Shuar	13%	74%

Source: Interviews with rural household heads: 1986 and 2011.

with the productivity of the pasture for the next year, the typical length of a rental contract, so they will eliminate all seedlings from the pasture in order to enhance the regrowth of the pasture in the following months. Owners, who operate with a longer time horizon, will identify the species of the sprouting tree. If it looks to be commercially valuable, they will allow it to grow in place, anticipating that ten to fifteen years later, they will be able to harvest the tree and sell it to a sawmill or to a furniture maker for a considerable sum of money. In other words, renters and owners have different discount rates. Renters have high rates, and owners have low rates. Because the seedlings appear spontaneously, the creation of these spontaneously generated silvopastures entails considerably less labor from farmers than most silvopastoral systems in which farmers plant the trees in the pasture.

As noted, these silvopastures have emerged more frequently in some communities than in other communities. In addition, they seem to have appeared recently, with the highest densities of trees occurring in communities that have a high proportion of recently germinated, relatively small trees. Higher stem densities are most likely to occur on farms where a "son changes the cows" and cleans the pastures.[2] Stem densities averaged 366 per hectare where the sons changed the cows and 200 where they did not ($p \leq 0.001$). The higher level of involvement of the younger generation in management of these silvopastures would be consistent with the observation that the silvopastures have emerged recently.

The ecological effects of these different patterns of pasture management by owners and renters can be considerable, as indicated by the data in table 5.3. It indicates the density of three types of tree seedlings in owner-occupied versus rented pastures. The left column indicates the density for

Table 5.3

Stem densities of tree seedlings in cattle pastures, Ecuadorian Amazon, 2011

	All stems: Densities per hectare in pastures	Small stems: Densities per hectare in pastures	Palm trees: Densities per hectare in pastures
Owner-occupied pastures	262.4	124.6	17.9
Rented pastures	119.8	81.7	9.2

Source: Interviews with rural household heads: 1986, 1997, and 2011.

all kinds of seedlings that achieve chest high size. The middle column indicates the density of just the youngest seedlings, and the right hand column indicates the density of palm tree seedlings. Palms are included as a separate category of tree because their thick trunks make them so difficult to chop down that farmers frequently leave them standing. The same pattern of regeneration runs across all three types of seedlings. Owner-occupied pastures have many more seedlings than rented pastures. All of the mean differences across the different types of seedlings are statistically significant.

The carbon-sequestering effects of these differences in seedling density can be significant. Calculations of the carbon sequestered, above and below ground, in the owner-occupied fields indicate an increment of about one ton of carbon per hectare per year compared with the carbon sequestered in rented pastures. At the July 2018 price of carbon offsets, the smallholders with high densities of trees in their pastures would receive $12 to $14 per hectare per year. The $200 to $300 annual increment in income from carbon sequestration would provide a noticeable increase in income for smallholders whose total annual income might approximate $5,000. In addition, the trees in pastures provide nutrient-rich litter for the pastures and protection for water sources for the cattle. Because the predominant pasture grass in the region, *gramalote* (*Axonopus scoparius*), is a shade-tolerant grass, the productivity of the pasture does not decline under trees. As a result, the stocking rates for cattle in shaded pastures was no different than the stocking rates for sun-filled pastures.

The fields with few trees show some signs of soil exhaustion. The pasture grasses mature more slowly than they did in the past, sometimes taking a full year before they flower rather than the eight months that it took when the fields were first planted with pasture grasses. Bare spots and inedible scrub growth have colonized sections of these fields. Because rental agreements typically last for only one year, the renters who pasture their cattle on these lands have no long-term investment in them and no incentive to address issues of pasture degradation. Taken together, the association of negative environmental outcomes with rented lands and positive environmental outcomes with owner-occupied lands seems supportive of Netting's argument about the affinity between owner-occupied lands and the practice of small-scale sustainable agriculture.

The patterns in these data also suggest trends associated with globalization, in particular that the growing inequality in landholdings contributes to environmental degradation at the lower end of the size distribution of landholdings. Owner-occupied landholdings in Morona Santiago averaged 25.2 hectares of pasture compared to 16.0 hectares of pastures on farms that rented pastures to others (differences in means, $p = .046$). As landholdings decline in size through subdivision or sale, landowners seem more likely to become passive owners, renting their land out to others and doing little to improve the long-term sustainability of their pastures. At the other end of the size distribution of landholdings, large operators can most easily increase the scale of their operations, and enjoy the advantages of increased scale by renting additional lands from other smallholders.

As noted, these patterns of cattle pasture management have a pronounced ethnic dimension. Table 5.4 compares the stem densities of trees in mestizo- and Shuar-owned pastures. Contrary to the stereotypical notions about the *ecological noble savage* (Redford 1991), the mestizos in Morona Santiago, not the Amerindian Shuar, practice the more sustainable regimen of pasture management. All of the mean differences in this table are statistically significant. The Shuar in fact are caught up in a natural-resource-degrading poverty trap. They are forced, because of a lack of palatable economic opportunities, to degrade their chief asset, their pasture lands, through rental contracts year after year in order to earn their subsistence. In this sense their poverty induces them to practice a strategy of pasture management that over time will diminish their returns from the land even further (Lerner et al. 2015; Rudel, Katan, and Horowitz 2013). A natural-resource-degrading poverty trap really represents an impoverished, rural form of environmental injustice that is meted out to rural peoples without enough wealth to escape the trap.

Table 5.4

Stem densities of tree seedlings in cattle pastures, Ecuadorian Amazon, 2011

	All stems: Densities per hectare in pastures	Small stems: Densities per hectare in pastures	Palm trees: Densities per hectare in pastures
Mestizo-owned	269.8	129.7	19.3
Shuar-owned	84.8	64.2	5.0

Source: Interviews with rural household heads: 1986, 1997, and 2011.

Conclusions: How General Is This Pattern? What Are Its Policy Implications?

What do these patterns tell us about the link between landownership and the ecology of small-scale, sustainable cattle ranching? The emergence of these silvopastoral landscapes on owner-occupied lands but not on rental lands suggests a logic that resembles Netting's (1993) premise about the sources of sustainability. It has occurred in a setting with two countervailing forces, a globalizing market for beef and a rise in part-time farming, so its presence in this setting testifies to the robustness of the connections among small-scale agricultural operations, landownership, and sustainable practices. The particulars of the sustainable practice may be important in assessing the significance of the Morona Santiago case. Most, but not all, sustainable agricultural practices are labor-intensive activities. Terracing lands, manuring fields, and maintaining home gardens all require additional inputs of labor when introduced. The conversion of a pastoral to a silvopastoral landscape usually follows this pattern because the landowner has to plant the trees, oftentimes along the edges of fields. In the case of the spontaneous silvopastoral landscapes of Morona Santiago the additional inputs of labor are insignificant. The person who cleans the pasture with a machete after the cattle have grazed it just leaves the seedlings of valuable tree species in place. The absence of additional labor inputs makes it possible to create this type of silvopastoral landscape even in places where the growth of non-farm employment opportunities in nearby towns has increased the opportunity costs of work on the farm. Because the trees create another income stream for landowners through the sale of wood to sawmills, their presence in pastures counteracts, to some degree, the episodes of extremely low prices for beef like those that occurred in the early 2000s when the Ecuadorian beef market experienced globalization and the entry of extremely inexpensive beef from overseas. In this sense the emergence of a silvopastoral landscape in cattle ranching areas makes these places more food secure by making small, local producers more resilient in the face of market volatility because they have more diverse sources of income.

The generality of a case is always open to question in assessing the policy implications of a case study. Pastures are the most common category of land use in the world, so enhancing the sustainability of pastoral land uses through measures like the ones described here should be a high priority in global efforts at sustainability. More particularly, are spontaneously generated silvopastoral landscapes a common enough occurrence in the tropics to warrant promotion through some sort of payments for environmental services (PES) program (Wunder 2005)? Silvopastures occur in a wide range of settings in the tropics, in Central America in Costa Rica (Harvey and Haber 1999) and in sub-Saharan African settings like Cameroon (Carriere et al. 2002). The extensive Babassu palm forests of the southeastern Amazon basin in Brazil spread spontaneously after deforestation and, typically, fires (Anderson, May, and Balick 1991). Interestingly, the presence of fires and their role in propagating the Babassu silvopasture contrasts with the humid, fire-absent silvopastures in the Ecuadorian Amazon. Certainly, the absence of fire in Morona-Santiago does not seem to be a condition that limits the generality of the lessons that one can draw from this case.

The ongoing globalization of the beef industry evident in the history of cattle ranching in Morona Santiago has several contradictory implications for sustainability initiatives in the cattle industry. Global trends in crop production over the past two decades underscore the growth in the production of oil seeds like soybeans and the relative stagnation in the production of basic grains like wheat. These trends testify to the growing influence of urban consumers and their growing demand for more animal protein, sometimes produced in very specific ways (Rueda and Lambin 2013). In this context Brazilian cattle ranchers have been able to secure a high-end market for '"grass fed" beef overseas, in Europe in particular. Cattle produced in silvopastures would appear to provide a similar opportunity in Ecuador. Certification schemes could focus, conceivably, on the sustainability of pasture management routines, a dimension along which the Ecuadorian silvopastures described here would score well. Entry into this type of high-end market for beef would provide a further financial supplement to PES schemes that could pay for sequestering carbon in silvopastures.

While globalization can promote sustainability through the growing influence that affluent, distant, and environmentally concerned consumers can play in shaping production practices (discussed in chapter 9), globalization has a contrary effect through its impact on landownership and rental practices in the Morona-Santiago region. The political activities during the 1970s and 1980s of Shuar and mestizo settlers, channeled by the agrarian reform and new land settlement legislation of the 1960s, created a relatively equitable, owner-occupied cattle ranching landscape by the end of the twentieth century. Globalization, with its destructive impact on local cattle herds, its promotion of growing inequalities in the size of farms and cattle herds, and its encouragement of land rentals, has undermined the social bases for silvopastoral agricultural sustainability among the cattle ranchers of the Ecuadorian Amazon.

To date, both the environmentally curative effects of the new silvopastures and the environmentally degrading effects of the continuous pasture rentals have occurred without meaningful interventions by either the state's agricultural extension agents or environmental NGO personnel. In this respect, Morona-Santiago represents a *rural backwater*, largely untouched by agricultural policy initiatives of any kind. The central government of Ecuador could provide an infusion of capital, but the political calculus that would lead a government to invest in a rural backwater like Morona Santiago does not seem self-evident. For that reason, there is little reason to believe that government assistance will be forthcoming for the Shuar smallholders with degraded lands or for the mestizo smallholders whose pastures merit carbon sequestration payments. In this sense, the dynamics described here have occurred in the absence of a concerted effort to improve Morona-Santiago's coupled human-and-natural system of cattle ranching through improved governance. It may take a large-scale, internationally funded effort to encourage the spread of these spontaneously generated silvopastures and arrest the degradation of the rented pastures. In sum, the currents of change in cattle ranching described here present opportunities for mobilization around new policies that would have important ramifications for global-scale issues like climate change and biodiversity conservation.

Notes

Funds from a Fulbright fellowship and two grants from the National Science Foundation (SBR9618371 and CNH10009499) facilitated this research. Delores Quesada Tenesaca, Gerardo Caivinagua, Amy Lerner, Diana Burbano, Megan McGroddy, Carlos Mena, Laura Schneider, Tuntiak Katan, Bruce Horowitz, Diane Bates, and Rafael Machinguashi provided very valuable help with this research project at different points in time. The good offices of the *Universidad San Francisco de Quito* and CIAT (*Centro Internacional de Agricultura Tropical*) also furthered this research.

1. "Pre-titles" are, as the name implies, not full titles to land. Smallholders can avoid the full expense of titling by acquiring only the pre-title, which is sufficient to transfer the ownership of land and to put the land up as collateral in order to obtain a bank loan.

2. People "change the cows" daily. Pastures are not fenced. Rather, cows are tethered in a patch of mature grass. Over the next twelve hours they eat all of the grass within reach. Then someone comes along and pulls the cow into another patch of mature grass before tying it down again. Water in the predominant pasture grass in the region is sufficient to keep the cows hydrated.

6 Cheap Meat and Cheap Work in the U.S. Poultry Industry: Race, Gender, and Immigration in Corporate Strategies to Shape Labor

Carrie Freshour

The global expansion of meat consumption is linked to the industrial production of meat, which relies on cheap inputs like feed grains (Winders 2017), the growth and consolidation of corporate power (see chapter 2), and the liberalization of international markets (Winders et al. 2016). This chapter focuses on another key feature of global meat production: the maintenance of a cheap, global labor force. While the U.S. poultry industry creates trillions of dollars in annual revenues ($63.9 trillion in 2015) and hundreds of thousands of jobs (281,000 in 2017) (Kay 2018), the industry relies on and must maintain a cheap workforce.

I make several arguments about labor in the U.S. poultry industry. First, the current labor conditions are reflective of and emerge from the historical conditions foundational to the industry. Second, there is a continuity in the reliance on marginalized groups as workers in poultry processing, even if these groups have changed over time. Third, the changes in which groups work in poultry processing are tied to dynamics of corporate practices and state policies. I demonstrate this third point by showing why poultry processing has recently shifted from relying heavily on undocumented immigrant women (Latinas) to reemploying African American women.[1] I conclude with a brief discussion on the need for both local and transnational social movement organizing, as the largest firms, Tyson, JBS, and WH Group, consolidate power and expand operations across national borders, as seen in chapter 2.

Global Implications of the U.S. Model

While this chapter focuses on work and workers within the U.S. poultry industry, its implications are reflective of accumulation strategies employed by large meatpacking and processing companies around the world. Vertical and horizontal integration methods continue to expand globally as a model for concentrating profits and fueling rising consumer demand (Boyd and Watts 1997; Constance et al. 2013; Patel and Moore 2018; Striffler 2005; Weinberg 2003). The Food and Agriculture Organization of the United Nations predicts that poultry consumption will increase at a rate of 2.3 times between 2010 and 2050, compared to beef and pork consumption, which is expected to increase between 1.4 and 1.8 times over the same period (Weis 2013a). While the United States remains the leader in poultry production globally, production in Brazil, China, Thailand, and Mexico has increased over the last two decades through similar practices of vertical and horizontal integration, consolidation, and reliance on a marginalized workforce (see Hodal 2016).

Raj Patel (2016) argues the current global transformation of meat production and processing follows the trajectory of other forms of industrialized agriculture through the displacement of local and small-scale farmers and butchers. For example, Tyson has expanded global production by opening up processing plants across the global south, currently employing 5,000 workers in processing plants across China and India (Tyson Foods 2018). Through global production, the largest transnational meat corporations also transform local production practices. As recently as 2012, 80 percent of the 10,000 poultry slaughterhouses in China were classified as "artisanal" or "un-mechanized," but large-scale slaughterhouses are rapidly increasing throughout the country (Gao 2012; see also chapter 4).

Despite the restructuring of Chinese poultry production, the Chinese state has recently placed an "anti-dumping" ban on Brazilian chicken in an effort to protect its domestic production from cheap imports (McDougal 2018). Yet most countries across the global south have less power in protecting local production. Consider the case of South African poultry; since 2015, the U.S. poultry industry has lobbied for favorable terms of trade through the African Growth and Opportunity Act (AGOA). Backed by Senators Isakson (R-GA) and Coons (D-DE), the U.S. and South African governments amended AGOA through the Paris Deal in 2015 (Isakson and Coons 2015). This agreement removed South Africa's anti-dumping tariff and increased the annual quota of U.S. poultry imports to 65,000 tons (Bavier 2018). Since implementation in 2016, South Africa's poultry industry has lost an estimated 5,000 jobs and is expected to lose a total of 110,000 jobs across the poultry sector in the coming years (*Africanews* 2017). As Kevin Lovell, president of the South African Poultry Association puts it, "We [South Africa consumers] have become a waste receptacle for the developed world" (Seeth 2017). The "off-cuts," of less desirable bone-in thighs and drumsticks, are dumped onto the global south creating a "cheap" product for urban consumers at the expense of local production.

While the economic value of corporate meat production has increased nearly sixfold in ten years (from 2004-2014) and is now valued at $366 billion (McDougal 2018), this global expansion has had devastating effects for the environment, consumers, and workers. With poultry expected to surpass beef and pork globally by 2020 (OECD/FAO 2014), the expansion of the industry requires a growing workforce. We can expect labor practices, following environmental standards, to worsen when moving from the United States to countries and regions with lower wages, less regulation, and a "comparative advantage" in the global "race to the bottom" (McMichael 2017). For example, Cargill recently opened a poultry farm and processing operation in China's Anhui Province, which is a province known for lower land and labor costs. The plant employs more than four thousand people and processes 225 birds per minute (bpm), which amounts to more than 65 million chickens per year (He 2013). These line speeds are considerably faster than those within the United States, currently capped at 140 bpm, yet provide fodder for industry-led movements within the United States to deregulate line speeds and remove caps altogether.

The Poultry Capital of the World

U.S. poultry production is firmly concentrated in the American South, and Georgia has had greater poultry production than any other state for the last thirty-six consecutive years (UGA CAES 2017). In 2015, this industry employed 29,831 workers in poultry processing alone (Georgia Power 2016). Production across the state has more than tripled since 1978 to meet exponential growth in per capita consumption across the United States and around the world. Yet, it is no coincidence that this self-declared "Poultry Capital of the World" also proudly boasts a "competitive advantage" for manufacturing through extremely low unionization rates (2.7 percent) and the lowest average of all hourly poultry processing wages ($9.54/hour) (Georgia Power 2016).

The scholarly literature (Gisolfi 2017; Gray 2014; Ribas 2015; Striffler 2005; Stuesse 2016) as well as popular and policy-oriented reports (Compa 2004; Fritzsche 2013; Oxfam America 2016) suggest that the industry employs mostly immigrant and largely undocumented workers from Mexico and Central America. Yet, through the interplay of state and corporate action, poultry processing labor is undergoing another major shift, back to a majority native-born black workforce. This shift reflects a longer history of the industry's dependence on a precarious and racialized workforce, often poorly subsidized through federal and nonprofit social service sectors (see chapter 2). In the following section, I outline the relationship of class struggle to the industry's changing workforce, as corporate strategies and state policies produce cheap labor to facilitate global expansion.

Class Struggle and the U.S. Poultry Industry

The history of the U.S. poultry industry is a history of class struggle. Up until the 1940s, poultry production remained very small scale, localized, and was considered "women's work" for household consumption and a meager extra income (Gray 2014). Yet, soon after the Great Depression, white landowning farmers and merchants in Northeast Georgia took over the industry and its profits. According to Gisolfi (2017), this takeover was structured along the preexisting crop lien system used in the overproduction of cotton that once dominated the region. Racially discriminatory state interventions under the Agricultural Adjustment Act (AAA) worked together with an agricultural credit system, creating structural barriers to commercial poultry production for black and poor white sharecroppers and tenant farmers.

The AAA programs subsidized cotton planters to idle land and quite literally displace farm labor (Raper [1936] 2005; Gisolfi 2017; Woods 2017). Between 1935 and 1940, the number of tenant farmers in the region dropped by almost 25 percent as planters destroyed cotton crops in exchange for AAA allotment checks (Gisolfi 2017, 13). This newly "freed" population was disciplined to become wage labor for the South's agrarian economy or for the industrial cities of the North, in turn transforming class structure in the South (Winders 2006). As a result, cotton planters lost considerable political power to the Midwest corn lobby to the ultimate detriment of supply management policy (Winders 2006). As the position of the U.S. cotton industry weakened, planters were forced to seek alternatives to cotton, creating the conditions for the expansion of the poultry industry. In effect, cash-poor farmers were transformed into "little more than hired hands" (Fite quoted in Weinberg 2003, 10).

While the poultry industry first flourished in the Delmarva peninsula, competition took off in the South due to the "pioneering" work of Gainesville, GA local, Jesse Jewell (NCC 2013) and leaders within the Georgia Cotton Producers Association like D. W. Brooks who turned from raising cotton to poultry (Dimsdale 1970). Jewell bundled baby chicks with feed to farmers. After they grew out the birds, he transported and slaughtered the chickens. The Georgia Cotton Producers Association, later renamed Gold Kist, played a similar role across much of the South, supplying feed, establishing a hatchery, and securing processing facilities. The industry received a major boost in 1944 when the War Food Administration reserved all the chicken produced from seven counties in North Georgia, ensuring a guaranteed buyer (Weinberg 2003). The corporate-state nexus was foundational to the growth of North Georgia's poultry industry, through a combination of federal loans, purchasing guarantees, and private industry integration. This relationship continues to benefit the largest meat producers (see chapter 2).

Marginalized workers were pivotal to the industrial transformation of the industry. Women came from cotton tenant farming and sharecropping livelihoods "out in the country" of the surrounding counties, considered by Jewell to be "uneducated," "unskilled," and desperate for waged work (U.S. House 1951). During Jim Crow, poultry processing was one of the few industries that hired black people. Faye Bush, who started at Georgia Broiler in 1948 at the age of sixteen, describes the relationship of Jim Crow to poultry hiring practices, "Well, it was the first public job that black peoples had to work on, so I thought it was good because I thought I was making some money."[2] She worked "ruffling" and "picking" chickens for fifteen years.

Gene Masters, a general manager of a plant in Albertville, Alabama, discusses hiring practices in the early days (1959–1964).

GM: The plant that I ran was ... we did a[n] extremely poor job of selecting employees. The education level was very, very low. We hired people directly off of the street when we needed them.[3]

Masters's comments suggest that the hiring process then was not vastly different from hiring practices today. Similar "just in time" hiring practices that mirror Masters's "directly off the street" and "when we needed them" practices remain. Prior to legal integration of workplaces, black workers would often only be brought in as scabs, and employers praised them for their "loyalty" and "willingness to do the dirty work which soon became distasteful to foreign women" (quoted in Horowitz 1997, 200–201). The poultry attracted black workers because it provided higher wages and social insurance programs from which they were excluded as agricultural and domestic workers (Glenn 2002; Hunter 1997; Winders 2009; Woods 2017). By the early 1960s, black women had largely displaced white women in Northeast Georgia (Gray 2014).

Large-scale unionization in the South often lagged behind efforts within industrial cities to the north. Yet, at Georgia Broiler in Gainesville, Faye Bush recalls a time in the early 1960s when a group of "mostly black people" walked off the line in protest against the heat. She looks back on the walkout both noting the fear that workers had, but also the deliberateness of the action, which she clarifies for the interviewer.

FB: I think it was just the fear of speaking out. Because I think if we could have spoke[n] out more, we could have done more. I think most people was afraid to speak out because of the way they had been all their lives. I remember one time the working condition was so bad, it was so hot in there 'til we found ourselves walking out. But they gave you a certain time to be back on the line. Then you would lose your job. Jobs didn't come by that easy back then. Peoples was afraid.

CW: So, are you saying that sometimes you would just walk out just to get some air?

FB: No, we called ourselves, "going to have a walk out!" Everybody walked out in order for them to get some air to make it cooler. So they put ice in there and then they gave us a certain time to come back to work.

CW: So, you walked out as a protest?

FB: Right, right.[4]

Similarly, Masters described labor discipline as a major problem in the early 1960s, at a time similar to Bush's walkout. In a plant of 240 workers, where 95 percent were black, Masters recalls workers regularly controlling their time through absenteeism. This was especially common on Mondays after a weekend of reprieve.

GM: Now, there were other problems, but for instance the owner of the company that I worked for told me one time, "I'm tired of all these employees coming in late, drunk and have a bad weekend and show up at noon on Mondays and so on. So, if any group doesn't show up at a certain time I want you to fire the group." That was the rule. Of course, when I did that and one of them was a floor lady that we needed the boss changed his mind [laughter]. ... Yeah, we did have serious problem with weekend hangovers.[5]

This practice of weekend-hangover-induced absenteeism mirrors historic patterns of worker resistance documented elsewhere as one in need of direct disciplining (Linebaugh 2003; Thompson 1967). This was a problem of labor discipline that the poultry, in its early days, could not control. While work was dirty and difficult, workers maintained some power vis-à-vis their employers, which increased in the coming decade through organizing by black workers across the South.

Black Worker-Led Organizing

The first documented large-scale poultry strike, which was led by "mostly black women," took place at Sanderson Farms in Laurel, Mississippi, in February 1979 and lasted until December 1980 (Schwartzman 2013). The strike focused on the pace of work, bathroom breaks, and sexual harassment by male foremen. Workers were "tired of being treated like dogs!" (Brown 1979). ICWU spokesman Bob Kasen explained the union's efforts as representing a shift in organizing strategy, "We've believed for some time that we ought to be functioning with people who get banged around the most—blacks and women" (Brown 1979).

This common knowledge of "getting banged around the most" is evident in workers' interviews and allowed for the construction of a shared working-class identity that often united workers across race for a time, in opposition to management. Don Mays, a black man who worked "live hang" in several Gainesville plants, commented on this change from the time he started working at the plant in the early 1970s.

DM: Okay, like I said, at first there were only a few Vietnamese there, a very few. It was mainly Black and White. And we got along, kind of great I guess, because of the fact of where we were. Everybody knew that people looked down upon us. ... They are glad you are there, but nobody else wants to do it. It was great coz it was kind of like a band that everybody, we all knew that we worked in poultry places, we knew that, "the undesirables" as we'd call each other, nobody else wanted to associate with us but we were all just one big family.[6]

Importantly, Don Mays distinguishes the "undesirables" from management because they were "gettin' their money."[7] In this sense, workers had a shared working-class consciousness.

National unions, like the ICWU, saw the necessity in organizing the South in the midst of deindustrialization and a decrease in real wages for most manufacturing jobs (Smothers 1996; Horowitz 1997). Unlike previous union drives in the South—such as the CIO-led "Operation Dixie" beginning in 1946, which brought in outside organizers and did little to challenge existing racial hierarchies—this labor movement practiced "social movement unionism" that centered on building coalitions with existing nonlabor organizations working for economic and social justice while centering black leadership and civil rights (Fantasia and Voss 2004; Schwartzman 2013; Seidman 1994). The goals of labor organizing expanded to address economic and racial justice head on.

From 1979 to 1995, there were at least eight documented strikes and walkouts in poultry processing plants across the U.S. South (see table 6.1). In North Georgia, there was at least one major walkout during this time period around contract negotiations: "Around 50 disgruntled night-shift employees at [the local] poultry processing plant walked off the job Wednesday at 5:45 p.m. without the support of their union. They vowed not to end the wildcat strike until they are given a $1-an-hour pay raise, better benefits and improved working conditions" (Ready 1978). Theresa, who was nineteen at the time, participated in the strike. Her mother worked at the plant for twenty-six years, and she started working on the evisceration line right after high school. She recalls the breaking point for a younger group of workers who would not be satisfied with the working conditions of her mother's generation.

Theresa: We got to the point where the conditions got so bad that we actually had a walkout. It was protesting ... I think by us doing that they recognized that, "hey, they had a reason to do that." You know, everybody

Table 6.1

Labor organizing and events across the Southern poultry processing industry

1951	Gainesville, GA	Jesse B. Jewell Inc. uses mob violence against AMC
1972	Forest, MS	60 workers walk off line and form Mississippi Poultry Workers Union
1978	Athens, GA	69 employees terminated for wildcat strike at Gold Kist plant
1978	Durham, NC	Amalgamated local 525 strikes for two months at Gold Kist plant, joined by the Progressive Labor Party
1979	Laurel, MS	211 out of 291 workers strike at Sanderson Farms
1982	Buena Vista, GA	22 black women walk out Buena Vista to join RWDSU
1988	Greensboro, NC	1,000 out of 1,140 wildcat strike at House of Raeford Farms Inc.
1990	Wilkes County, NC	70 drivers at Holy Farms strike and join the Teamsters
1991	Hamlet, NC	Imperial foods fire kills 25, injures 56, out of 245 workers

felt if we did something then, my mother they worked under conditions way worse than I did. Now the women they have equipment, they have it better.[8]

The most vocal leader of the strike was Fred Faust, a knife sharpener who had worked at the plant for four years (Ready 1978). Faust put workers' complaints clearly: "We haven't had a raise in over a year. They just don't respect us. They want everybody to do two jobs" (Ready 1978). These complaints reflected the speedups taking place across the industry. It was over this time period that the poultry industry experienced growth in further processing—so much so, the industry's workforce grew by almost 96 percent between 1972 and 1992 (Schwartzman 2013, 41).

As growing numbers of Southern black women organized, state policies and corporate strategies combined to undermine their efforts. First, the federal government became increasingly anti-union, as seen in President Reagan's response to the PATCO strike in 1981 effectively removing poultry processing workers' most powerful weapon to date, the wildcat strike. Second, the passage of NAFTA during the Clinton administration facilitated corporate relocation strategies that transformed the poultry industry by opening Mexican markets to U.S. poultry production (Bacon 2012). And third, as labor lost power, the U.S. poultry industry lobbied to weaken federal line speed regulation, increasing production limits from 35 birds per minute (bpm) in 1970 to 91 bpm in 1990 (Albert 1991). These speedups increased the production of cheap processed chicken and by 2000, 90 percent of chicken was sold in pieces (Fink 2003, 12; Patel and Moore 2018; Simon 2017; Striffler 2005). These three seemingly separate events represent class struggle within the poultry industry, which effectively weakened black labor, increased turnover, and opened the door for industry's recruitment of undocumented immigrants. In this sense, labor conflict within the United States laid the foundation for the global expansion of the meat industry, as workers lost power vis-à-vis corporate and state actors.

Labor Displacement through the "Hispanic Project"
Between 1990 and 2010, the "Poultry Capital of the World" saw its Hispanic population increase by 683 percent (Ennis, Rios-Vargas, and Alber 2011, 6; Bureau of the Census, U.S. Department of Commerce 1990, 16). Georgia tied North Carolina as the top immigrant-receiving state during this time.[9] Both Dalton, Georgia, as "the carpet capital of the world," and Atlanta, in preparation for the 1996 Olympics, attracted Latina/o immigrants, most notably from Mexico, and Latina/o migrant workers from the U.S. West coast (Odem 2009; Zúñiga and Hernández-León 2005). Yet, Latina/o workers were a small part of the poultry industry workforce until the mid-1990s (Kandel and Parrado 2005; Marrow 2011). Yet, by the 2000s, they made up 75 percent (Fink 2003; Griffith 1990; Guthey 2001; Striffler 2005). They did not simply show up, but were actively recruited in the 1990s by the largest corporations in the U.S. poultry industry—Tyson, Pilgrim's, and Gold Kist. I outline this transformation using ethnographic research conducted in and around one of the largest plants in the state. This plant, which will be referred to as "Acme Chicken Processing" (ACP), employs over 1,200 workers, and is owned by one of the top five processors in the world.

Aida, a librarian and local immigrant-rights activist in Northeast Georgia describes how the first immigrants came to work in one of the state's largest plant in the early 1990s:

Aida: Yes, we would ask, "How did you came here?" They all said that when they were coming over from their countries, usually through Mexico, there were signs, like in Texas that says "Jeo-rja poultry is looking for employees." So, they came up, and it took a few to get here and then they would tell their family members, so a lot of people came over.[10]

Across the broader region of Northeast Georgia surrounding ACP, I documented many anecdotal accounts of the plant recruiting workers at the flea markets and Sunday soccer leagues, and even distributing false documents from ACP's HR (human resources) office. These allegations were confirmed in 2007 when two HR employees won a lawsuit against ACP, for discrimination and wrongful termination. One of ACP's HR employees was found guilty of making fake Social Security cards for undocumented workers (*Reyna v. ConAgra Foods, Inc.* 2007). Across the South, the largest poultry companies cheapened labor costs by "bringing the third world in," in turn facilitating the expansion of global production.[11]

On a global scale, neoliberal deregulation of international markets beginning in the 1970s created the conditions for the U.S. poultry industry to grow exponentially. Subsequently, Mexico oriented its agricultural production for export, dispossessing farmers from basic food production. Tyson aided this process in 1989, when the company expanded production to Mexico, partnering with Trasgo and soon after opening Tyson de Mexico. NAFTA only exacerbated this unequal trade relationship, undermining local production and "freeing" Mexican labor to join international migration streams. Companies like Tyson recruited workers with the promise of permanent employment, higher wages, and housing (Fink 2003; Griffith 1990; Mize and Swords 2010).[12]

Black workers witnessed these changes firsthand. Don Mays commented on this process of labor displacement, "the Hispanics, they were just takin' the jobs that was given."[13] Mays goes on to note that everyone—not just the new "Hispanic" workers—worked hard at the poultry. He believed that the difference was that these new workers did not complain.

DM: They would work hard, just like anybody else. ... And it got to the point where they would work, they did real good, they would never complain, they'd just came in, and did their work and went home. As to where a lot of us would might complain about something we didn't like, about safety things, the raise, the way we was being treated, the way that inspectors was acting, or just any little bitty thing. But then, like I said, it seemed like there came a time when you started voicing your opinion about something a couple of, maybe a week or two later, something may come up and you're not there anymore. You weren't replaced with another American, you was replaced with a Hispanic.[14]

The cheapened workforce initially served the poultry industry well, replacing the most outspoken workers. Union membership at ACP dropped from 80 percent to 40 percent (Aued 2007). But as undocumented migrants began settling throughout the South, they not only became rooted with families in their new communities, but also began to expect and fight for their rights as workers and as human beings.

As early as 2006, the union local at ACP collaborated with another community organization, the Economic Justice Coalition (EJC), on a unionization drive at the plant. Linda Lloyd, director of the EJC, recalled the organizing drive: "The idea was to have a big cookout in both the communities, the Black community over at Riverbank and the Hispanic community out in Evergreen."[15] This collaboration paid off as the union local won the election in September 2006. These groups teamed up again a year later for the EJC's annual Labor Day march involving representatives from the local chapter of the NAACP and faith-based community, labor, and immigrant rights organizations. Across the South, coalitions among black and Latina/o workers were gaining strength (Bacon 2012; Stuesse 2016; Zepeda-Millán 2017). Undocumented Latina/o immigrants began making demands for union recognition and labor rights as well as for immigration reform, DACA (Deferred Action for Childhood Arrivals) and DAPA (Deferred Action for Parents of Americans), the right to drive, fair housing, health care, and transportation.

Nationally, the largest mass mobilization of undocumented immigrants occurred on May 1, 2006, when hundreds of thousands of immigrants committed to a general strike. This strike hit the poultry industry especially hard. Tyson, Perdue, and Gold Kist were forced to close 22 plants across the South. However, four months later, Immigration Customs Enforcement (ICE) agents conducted several workplace raids across the country. ICE is the federal agency used to enforce immigration law. These were the first of a series of raids, which disrupted poultry-processing labor once again.

The Gendered Racial Removal Program

In April 2008, ICE raided several of the largest plants across the South, arresting four hundred hourly line workers (Associated Press 2008). Pilgrim's Pride released an official statement a few days later emphasizing the company's use of E-Verify[16] and ICE's "Best Hiring Practices," working for the common goal of "eliminating the hiring or employment of unauthorized workers" (Pilgrim's Pride Corporation 2008). These raids constitute what Golash-Boza and Hondagneu-Sotelo (2013) call the "gendered racial removal program." First, the raids punish workers for organizing while disciplining the broader undocumented immigrant communities who remain living and working in the United States. Second, the raids provide state-directed labor displacement, forcing plants to find new workers.

The majority of the Latina/o workers left ACP or were fired at the end of 2008 and early 2009, and the plant returned to a majority black workforce. Workers like Alessandra, currently employed through a Temporary Protective Status (TPS) work permit, recalls people leaving every day: "They took them [undocumented workers] to the office and they tell them, 'so we need your birth certificate, and your social. If you don't have those papers with you, please don't come back.' We cry a lot."[17] This personal account was repeated time and time again and was not limited to ACP. Tom Frischitze, a labor attorney, was working in Alabama around this time for the Immigrant Justice branch of the Southern Poverty Law Center. He witnessed a similar practice in 2008 when plants routinely fired batches of workers, 30 to 40 at a time.

TF: The workers felt like it was clear the company knew they were employing a large number of undocumented workers, but if they actually verified everyone at once and laid off everyone at once who was undocumented they wouldn't be able to replace them. So, they were intentionally doing it on this rolling basis, so it would give the employer to find workers with authorization to replace undocumented workers in smaller groups.[18]

During this time, workers also left as anti-immigrant laws were passed across the state of Georgia and in other southern states. Senate Bill 350, passed in 2008, greatly limited the movement of undocumented immigrants by making driving without a license more than twice within a five-year period a felony. Activists in Northeast Georgia call this a "D.W.B" (Driving While Brown) and argue that this law increased the rate of racial profiling outside of their neighborhoods and the public schools.[19] In 2009, many counties across Georgia enrolled in Section 287(g) program of the Immigration and Nationality Act, which partnered local police departments with federal immigration and detention agencies. Georgia notoriously passed House Bill (HB) 87 in 2011, which requires all public and private sector employers with more than 10 employees to use E-Verify to determine their employees' immigration status and requires proof of citizenship to receive any public benefits. Thus, state legislation and industry compliance led to the transformation of poultry processing labor, back to a majority black yet disorganized workforce.

A few undocumented workers returned to ACP after the chaos of the raids and HB 87 died down. These workers rely on fake documents, and often their helmets display a name that is different than their own. They also lose all of their previously earned benefits. If the history of this industry teaches us anything, it is that the firing of undocumented workers and a return to a majority black workforce was a response both to growing visibility, dissent, and organizing among an undocumented immigrant population, *nationwide*, which worked alongside expanding *anti-immigrant legislation* across much of the American South. Yet, most workers who left ACP in 2008 and 2009 just moved to a smaller, less regulated, non-unionized plant about half an hour away. For the few Latina/o immigrants who remained at ACP on temporary work permits or maneuvering other precarious forms of documentation, the work has only become more difficult with faster line speeds, a decimated union, and even higher turnover rates. The risk and fear experienced by undocumented workers who stay in the plants is mirrored throughout their communities as thousands of undocumented immigrants continue to live in the poultry towns that brought them here. These laws increase precarity for undocumented immigrants while doing little to improve the conditions of the native-born working poor who remain.

Today, 80 to 90 percent of the line workers at ACP are native-born black women. While black workers have dominated this workforce in the past, this current generation of workers comes to the plants with severely weakened union representation or none at all. They carry the added fear of displacement either through plant maneuvering to return to undocumented workers or by the looming threat of offshoring. Even as hearsay, these threats kept ACP workers from participating in the kinds of labor struggles their grandmothers led decades prior. In this sense, labor disorganization produces "third world" labor conditions, even if undocumented workers are no longer employed. The surplus value produced across the American South bolsters industrial poultry expansion throughout the global south. In the next section, I will connect local and global labor struggles through a discussion of changing line speed legislation.

Local and Global Labor Struggles: The Case of Line Speedups

ACP runs two "sides" of operation: direct slaughter, commonly referred to as the "kill side," and further processing, or "debone," for fast food companies like KFC and Zaxby's. Twelve lines operate on the kill side. From here, much of the work is automated until it reaches the evisceration room. In this room, the sound of the machines is so loud that workers can barely hear the person beside them speak. Birds move on shackles along two main lines, each with six stations. The first worker receiving the bird is the "eviscerator." Although this position has been largely automated, the machines are never perfect, and it takes time to master. Every motion must be perfected to ensure efficiency (two to three seconds) or the birds will pile up and inspectors will have to stop the line. The actual prescribed process is rarely followed, as workers know they would not be able to keep up. Line speeds, then, shape workers' experience of the working day from the pace and pain of work, the length of each day, and whether or not workers must give up their weekends to the plant.

Workers generally do not see, but can feel on the line, the underlying ways in which federal agencies and the poultry industry advocate for line speedups, effectively disciplining workers inside the plant. Poultry processing line speeds are regulated by the USDA Food Safety Inspection Services (FSIS). As line speeds increase, the industry effectively creates what Marx calls "surplus value" by stealing time from workers, subsidized through workers' bodies and federal disability.[20] Don Tyson, in a *New York Times* interview, promoted further processing as "selling time"[21] (Frantz 1994). This "selling" points to an unequal valuation of time and people under capitalism as speedups in poultry processing are relational to changes in consumption and the valuation of workers' time outside of work.

From 2017 to 2018, poultry processing line speeds were a matter of national concern, yet driven largely by global production pressures. To briefly summarize, in fall 2017, the National Chicken Council (NCC) petitioned to increase the allowable maximum number of birds slaughtered per minute, from what they call "arbitrary line speed limitations" of 140 to 175 bpm, or preferably removing the cap altogether (Brown 2017). The NCC advocated for an increase in order to "level the playing field" by "eliminat[ing] competitive barriers between the U.S. and international chicken producers" (Brown 2017, 2) and "encourag[ing] more plants to participate in the New Poultry Inspection System" (13). In January 2018, the USDA FSIS denied the petition due to pushback from consumer and worker advocacy organizations (Rottenberg 2018). Opponents of the increase cited a host of studies, governmental and advocacy-led, that connect current line speeds to high rates of injury and illness, particularly carpel tunnel syndrome and other musculoskeletal disorders among poultry processing workers (Barnes and Morris 2016; Fortson and Hawkins 2015; Fritzsche 2013; Musolin et al. 2014; Oxfam America 2016; U.S. GAO 2017).

While this appears to be a victory for workers, taking a longer view of speedups within the industry reveals a troubling picture. Industry advocates and some USDA representatives have pushed for an increase to 175 bpm since as early as 1997 through the HIMP pilot, an acronym within an acronym, which stands for HACCP (Hazard Analysis and Critical Control Points-Based) Inspection Models Project. HIMP granted line speed waivers to 20 plants that allow them to operate at speeds up to 175 bpm (USDA FSIS 2015).[22] Each attempt to increase line speeds has been smartly packaged to improved technology and a more "modernized" scientific approach to biological hazards, food safety, and inspection, while ignoring not only worker safety but also the political struggles of worker organizing.

Thus, (de)regulation of USDA FSIS becomes another instrument used, in Karl Marx's ([1867] 1992, 261) words, to "shorten the part of the working day in which the worker works for himself," and "lengthen the other part, the part he gives to the capitalist for nothing." Not only are workers disciplined by the pace of work, but even after their shift ends, workers' "free time" is hardly free, spent recuperating for the next day, buying aspirin and gels at the dollar store. Their very lives are sped up, with a majority of workers experiencing "premature disability,"[23] in which they must piece together a living from a monthly disability check. In this sense, FSIS inspection joins a host of federal and industry-instituted policies to aid the exploitation of the poultry processing workforce.

Conclusion

Counter to popular discourse, for many of the most disenfranchised populations across the American South, "the poultry" provides an essential and even desirable form of "high" low-wage employment. Yet, the conditions of this work within the United States degrade as the industry expands production globally, competing with "cheaper" labor that the United States, in many ways, helped produce. Once, workers could use absenteeism and high turnover rates to gain some autonomy by moving in and out of a variety of low-wage jobs. Poultry plant workers with little control over the labor process would strategically take temporary pay cuts to provide brief reprieve for their bodies and momentary dignity for their souls (Griffith 1993). In the early- to mid-1990s, however, the largest plants used undocumented workers to undermine black worker organizing, and then easily escaped responsibility for these practices. Combined with disappearing social supports and increased policing and criminalization for a population with a historically tenuous relationship with the so-called welfare state, there is less dignity or choice involved in black women's decision-making to move in and out of this work. Additionally, the "gendered racial removal program" of undocumented immigrant deportation only increases the precarity of undocumented workers because most do not leave, but instead move to less-regulated plants.

The major demographic changes in the poultry processing industry over the past several decades have not simply happened, but are historically and globally shaped by class struggle. Tracing the production of cheap meat, from the perspective of labor, illuminates the ways in which this industry both depends on and maintains precarity for its low-wage workforce with lessons for the expansion of global meat production as state and corporate interests work to disorganize and displace labor. Through this framing, a cheap and constant gendered and racialized workforce is as integral to global meat production as the acres of GMO feed or the selectively bred broilers.

Contemporary struggles over line speeds are inherently global struggles in which worker advocacy groups must contend with National Chicken Council lobbyists' justifying line speedups by citing the threat of Chinese production. Yet as I have shown, speedups outside of the United States have been shaped by U.S. expansion, consolidation, and "dumping" practices globally. Therefore, labor movements must act globally, building coalitions not only among native-born and immigrant workers as in the case of the United States, but also with movements of workers in places like South Africa and China who seek control over their working day alongside antidumping campaigns for more just terms of global trade and the protection of local production. While the future for workers in this "meatification" appears bleak, labor movements in unlikely places such as the rural and racially and economically segregated South present a historic record of class struggle, one in which workers sometimes win.

Notes

1. This chapter draws on two years of ethnographic research in Northeast Georgia, between September 2014 and August 2016. I worked in one of the largest plants in the state, referred to in this chapter as "Acme Chicken Processing" (ACP), from November 2014 through April 2015, clocking in over a thousand hours of work on the line. This plant, typical of large poultry processors, employs over 1,200 workers, and is owned by one of the top five producers in the country. It is located in the Northeast region of the state, which has a long history of poultry growing and processing. I also rely on semi-structured, recorded, and transcribed oral histories with fifty-sixwomen workers, as well as interviews with thirty-onecommunity members, self-identified activists, educators, and political leaders connected to the poultry processing workers. Additionally, I draw on worker interviews conducted between 2000 and 2002 by Dr. Carl Weinberg and his students in and around Gainesville, Georgia. I supplement both sets of workers' accounts with newspaper articles and archival data collected in the poultry science collections at the Russell Special Collections Library at the University of Georgia.

2. Faye Bush, interview by Carl Weinberg, June 11, 2002.

3. Gene Masters, interview by Robyn McClure, April 27, 2000.

4. Bush, interview.

5. Masters, interview.

6. Don Mays, interview by Carl Weinberg, April 20, 2000.

7. Mays, interview.

8. Theresa, interview by the author, March 14, 2016.

9. The Pew Research Center reports that 58 percent of the fastest-growing Hispanic counties between 2000 and 2007 were in the South (Fry 2008).

10. Aida, interview with the author, January 28, 2016.

11. In 2001, six managers at a Tyson poultry plant were indicted with conspiracy to smuggle undocumented workers into the United States and knowingly employ them illegally. Yet, Tyson and three managers were acquitted in 2003 (Day 2003). In 2009, ICE agents raided Pilgrim's Pride plants in Batesville, Arkansas; Chattanooga, Tennessee; Live Oak, Florida; Moorefield, West Virginia; and Mt. Pleasant, Texas and detained 400 employees (Pilgrim's Pride Corporation 2008). The company reached a settlement the following year, agreeing to pay $4.5 million to a law enforcement fund at the U.S. Department of the Treasury and improve hiring practices. Pilgrim's also released a public statement denying all guilt (Garay 2009).

12 David Bacon (2012) documents a similar transformation in the pork industry through recruitment to Tar Heel, North Carolina, where former Mexican pig farmers became pork processing workers in the Smithfield Plant.

13. Mays 2000.

14. Ibid.

15. Linda Lloyd, interview with author, April 13, 2016.

16. E-verify provides an electronic service that matches worker information to records in the Department of Homeland Security and the Social Security Administration in order to verify employment eligibility.

17. Alessandra, interview with author, January 31, 2016.

18. Tom Frischitze, interview with the author, January 22, 2016.

19. In a recent survey conducted by UGA's Latin American and Caribbean Studies Institute (LACSI), 75 percent of the Latina/o community in the region drive with fear and do not drive more than is necessary (Calva 2016).

20. Social Security Disability Insurance (SSDI) and Supplemental Security Income (SSI) are both commonly referred to as "disability." SSDI provides long-term disability payments and is tied to having had employment. SSI provides short-term disability for adults and children with limited income and resources. Both require medical records to prove eligibility. Disability is only accessible for those U.S. citizens who persist in the application process. Currently, in Georgia, the average wait time for an SSI or SSD hearing is 16.6 months. The average case-processing time in Georgia is 575 days. The Georgia average for winning a disability hearing is 48 percent (Georgia Office of Disability Adjudication and Review 2018).

21. In 1991, chicken surpassed beef as the most highly consumed animal protein in the United States. This shift was based on Tyson's production model, and by 1995 95 percent of Tyson sales were further processed chicken products rather than the whole broiler (Kleinfield 1984; NCC 2018).

22. Since 2014, HIMP evolved into the New Poultry Inspection System (NPIS) with line speeds capped at 140, yet the twenty pilot plants continue to operate at up to 175.

23. Here, I am drawing on geographer Ruth Wilson Gilmore's (2006, 28) work on the political economy of mass incarceration, which she argues depends on an understanding of racism as "the state-sanctioned or extralegal production and exploitation of group-differentiated vulnerability to premature death."

III Consequences and Considerations

Up to this point, the chapters in this book have provided an overview of trends in terrestrial and aquatic meat production and case studies that seek to show how the growth of meat production has been shaped by broader structures, particularly state and economic policies and practices. These case studies provide much needed empirical evidence for why the growth of meat production has not been random or simply compelled by "natural" tendencies, but rather is the consequence of national policies and corporate strategies (chapters 2 and 4). In doing so, some chapters have also provided empirical evidence of the negative consequences of the globalization of meat, such as the treatment of workers in processing facilities (chapter 6) and the unintended environmental consequences of the increasingly global trade in meat (chapter 5). The concluding part III of this book expands our understanding of the environmental and social consequences of meat production, continuing the theme of much of this book, which is bringing empirical data to bear on our exploration of the global meat industry. In providing empirical evidence, the goal is to not simply describe the situation, but to also shed light on what needs to change.

Emancipatory Empiricism

We have set out in this work to offer evidence about the global meat industry with the intent of providing "emancipatory empiricism" (Jakubek and Wood 2018). What we mean by this is the use of systematic social science research methods, such as observations and surveys, to counter stereotypes and cultural biases. Using empirical data for emancipatory purposes allows us to "more accurately describe the relationship between social structure, agency, and the limitations that extralocal forces" place upon individuals' or communities' abilities to act (Jakubek and Wood 2018, 31). Moreover, in collecting empirical data about social life, in this instance the operation of the meat industry in diverse locations, creates an opportunity to imagine what is possible in the future.

The case studies in this book, in combination with the chapters in part III, speak to the relationship between social structure and agency, and the limits of agency within particular spaces. Specifically, each chapter provides empirical data that begins to fill in the outline of what exactly the "global meat" industry looks like, calling attention to the similarities across spaces (e.g., labor practices in slaughter plants), but also the differences (e.g., diverse state subsidies that have supported corporate meat companies in different ways). Providing these points of data is important for better understanding how policies and practices contribute to the problems we see today, but such data also can guide us in thinking more critically about the solutions. In other words, understanding the operation of the global meat industry in different geographical spaces and at different scales offers a clearer understanding of the growth of meat production, as opposed to simply accepting stereotypes and assumptions about the place of meat in our food system in the twenty-first century. Additionally, we have the ability to think critically about the future and what policies and practices would need to be put in place to create a different type of food system.

In terms of the problems in our modern global meat industry, chapters 7 and 8 shine a light on two of the more damning assertions made against the modern global meat industry today: greenhouse gas (GHG) emissions (chapter 7), and the treatment of animals within the system (chapter 8). These two issues are embedded in the very functioning of the intensification of meat production around the globe. Yet, both of these issues are hard to see on a daily basis.

In the case of climate change, empirical data is required to reveal the effects (e.g., longitudinal data of melting polar ice sheets, rainfall, temperatures). In the case of animal ethics, scholars have argued that intensive meat systems are designed to not be seen (Pachirat 2013). The lack of visibility surrounding animal and human suffering in CAFOs, feedlots, and slaughter houses occurs both because of the distance of most of these spaces from largely urban and peri-urban consumers and because the system is designed to resemble any other type of factory with assembly line production. The factory-like system ensures a division of labor and guarantees monotonous jobs, which means even those who do have contact with intensive production systems (workers, managers, local community members) can compartmentalize and get on with daily life (oftentimes out of necessity) without seeing the suffering inflicted upon animals as ethically problematic (see chapter 6; Pachirat 2013).

Take for example, live animal transport and how this practice ignores animal ethics and humans' obligations toward animals. Among the more inhumane forms of industrial animal agriculture, the practice has only continued to expand, with global trade in live animals "having grown from approximately US$7 billion in 2000 to more than US$19 billion in 2013 (the most recent year for which comprehensive data are available), an increase of almost 300 percent over a period of a decade and a half" (Keogh and Day 2016, 6). Most countries participate in the trade of live animals, but the Australian trade has received the most attention in recent years. Australia is one of the largest exporters of live sheep, the bulk of which go to the Middle East. The journey takes approximately three weeks, and the sheep are packed into poorly ventilated spaces, where temperatures can soar above 100 degrees on the ships during the summer (Glover 2018). These ships generally carry over 50,000 sheep, thus an accepted industry standard for the number of sheep that will die during transport is approximately 1,000, but there are documented instances of much higher numbers of deaths, including 4,000 in 2014 and 2,400 in 2017 (Foster 2018; Towie 2014). These reports of much higher deaths have led some to refer to these as "death ships" (Wright and Muzzatti 2007). Of course, the broader question that must be raised is how we came to a point in history where 1,000 sheep dying in transport is considered "normal" or business as usual.

Generally, trade in live animals is viewed as having limited oversight in the actual movement of the animals (as opposed to the extensive paperwork that is filed for the shipments), but trade has grown due to decreasing barriers to trade over the past two decades. Thus, international trade policies have played a role in increasing inhumane practices toward animals. Whether it is the issue of animal ethics and humans' obligations towards animals, or the impact that meat production has on GHG emissions and climate change, both issues require that we question the structure and place of meat in our food system today and in the future. Our final chapter (chapter 9) will take up the task of thinking about the future and current efforts underway to create change in our food system.

7 Contributions to Global Climate Change: A Cross-National Analysis of Greenhouse Gas Emissions from Meat Production

Riva C. H. Denny

Global climate change is already having measurable effects on sea level and global temperature averages (IPCC 2013), and animal agriculture is a substantial source of the greenhouse gas (GHG) emissions contributing to climate change. In 2015 animal agriculture contributed 65 percent of the total global emissions from agriculture and 7 percent of total GHG emissions globally (author calculations using FAO data and global data from Olivier, Schure, and Peters [2017]). Agriculture as a whole is responsible for about 11 percent of world GHG emissions. Although neither agriculture nor animal agriculture is the largest producer of GHG emissions, both are significant sources, and global meat production has increased (see chapter 1).

While there are many arguments against eating meat, it seems unrealistic to expect that the entire world population will give up eating meat in the near future (see chapters 8 and 9). Meat eating has strong cultural traditions, is an important source of fat and protein for many people living in marginal environments (grasslands, etc.), and is also a by-product of dairy, fiber production, and draft power. If we take as a given that meat consumption will continue at some level (not necessarily the current one), then the question is how and where should this meat be produced? Reducing the amount of GHG emissions that are produced by meat production is a step toward more sustainably feeding the world population.

GHG emissions from meat production are a particularly important type of environmental impact to consider from a social and environmental justice standpoint because the effects are ultimately felt on a global scale through global climate change. However, the effects of climate change are not being felt evenly (Hallegatte et al. 2016), and, as is the case with other sources of GHG emissions (like from fossil fuels), the places producing the most GHG emissions from meat production tend to be the same places that will be least affected by climate change or that have the most resources available to adapt to it, or both (see chapter 1).

This chapter considers the total GHG emissions and emissions intensities of beef, pork, and chicken production across countries and over time. What factors drive total emissions and emissions intensities? What is the relationship between total emissions and emissions intensities? Geographically, where are the most emissions coming from and where is the most GHG-efficient meat production? Does more intensive meat production produce fewer GHG emissions? GHG efficiency is one of several kinds of efficiency that should be considered when assessing the sustainability of meat production, but GHG efficiency has the greatest direct implications for global climate change.

Animal Agriculture and GHG Emissions

The main types of GHG emissions from animals are methane (CH_4), which has a global warming potential (GWP) that is twenty-five times that of CO_2, and nitrous oxide (N_2O), which has a GWP that is nearly three hundred times that of CO_2. These emissions are the result of enteric fermentation (i.e., a key process of ruminant[1] digestion) and from manure, which is either collected in pits or lagoons (e.g., "manure management systems"), left to decompose on a pasture, or applied to soils as a fertilizer. Animal diet (and more technical chemical manipulation of the diet) can influence emissions from enteric fermentation, with greater roughage/less nutrient-dense foods in the diet generating more GHG emissions than diets with higher amounts of concentrated feed (Macleod et al. 2013, xix). The emissions from manure management practices vary depending on their specifics and the temperature of the location, as do the emissions from manure left on pasture and applied to soils.

Worldwide, total GHG emissions from cattle, pigs, and chickens combined increased by 66 percent from 1961 to 2015. However, this increase in emissions has not been even across the globe. Emissions over the past fifteen years have increased the most in South America and Africa, and South America has been the top-emitting continent since 2001 when its emissions surpassed those of Asia. In 2015, the top three continents for total meat animal emissions were South America (558,000 gigagrams [Gg]

CO_2eq—GHG emissions converted into equivalent about of CO_2 based on their GWP), Asia (480,000 Gg CO_2eq), and Africa (297,000 Gg CO_2eq).

From 1961 to 2015, missions from meat animals increased from all sources. During this period, emissions from enteric fermentation increased 58 percent, emissions from manure left on pasture increased 92 percent, and emissions from manure management and from manure applied to soil increased 68 percent and 66 percent, respectively. These increases are not really surprising as meat production has increased during this period as discussed in the introduction and shown in figure 1.2. The portions of total emissions from each source have been relatively stable over time.[2] In 2015, the majority of total GHG emissions from meat animals came from enteric fermentation (60 percent), 23 percent came from manure left on pasture, 12 percent from manure management, and 5 percent from manure applied to soils.

The top graph in figure 7.1 shows that total GHG emissions from cows, pigs, and chickens each increased significantly from 1961 to 2015, as the global meat industry expanded. The overall increase in emissions during this time period has been greatest from chickens (a 683 percent increase), while emissions from pork have increased by 107 percent and emissions from beef have increased by 58 percent.

Overall, beef cattle produce the most total emissions (note the different scale for beef in figure 7.1), largely as a result of their physiology—they are large, relatively slow-growing ruminants and thus per animal produce the most GHGs (Gerber et al. 2015), even though beef production and the number of cattle slaughtered is comparatively low (see figures 1.2 and 1.3). Pigs produce the second-most emissions, followed by chickens. Of particular note, pork production in 2015 was about 25 percent greater than that of chicken (see figure 1.2), but pigs produced about three times more emissions than did chickens. This is also largely a result of size and physiological difference between pigs and chickens and the differences that result in growth rate, kind of food consumed, and digestive process and the GHGs that result (Gerber et al. 2015).

Emissions intensity (EI) is the GHG emissions produced divided by the amount of meat produced. Consistent with the total emissions for each type of meat, beef has the highest EI, then pork, and then chicken. In 2015 a globally average kg of pork produced more than twice as much CO_2eq GHGs than chicken, and a kg of beef produced about forty-four times as

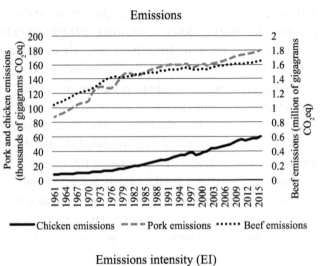
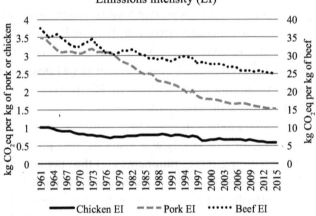

Figure 7.1
Total GHG emissions and emissions intensities (EI) for chicken, pork, and beef, 1961–2015
Source: FAO 2019.

many emissions as chicken and about sixteen times as many as pork (note the different scale for beef in figure 7.1). Not only does EI allow for comparisons between the emissions from different types of meat, it also lets us make some comparisons between places in terms of GHG efficiency.[3]

The bottom graph in figure 7.1 shows that the global EI for beef, pork, and chicken has decreased over time. While EIs have decreased over time across all continents as well (and most dramatically in Asia[4]), which would seem to be a good thing, much of this efficiency gain in GHG emissions has been achieved by raising animals more intensively, meaning more animals in less space, and getting them to grow as quickly as possible. This model of animal production creates concerns for animal welfare and environmental quality in the surrounding area due to dust, odors, and large volumes of animal wastes, as well as the potential for disease spread (and greater use of pharmaceuticals to control disease). This type of efficiency is also made possible by growing crops (like corn/maize and soybeans) specifically for animal feed, which creates other environmental concerns and GHG emissions that are not included in meat GHG emissions measures.

Figure 7.2 shows the 2011–2015 average beef emissions, beef production, number of cattle slaughtered, and EIs by continent and is ordered by total beef emissions. This graph and the ones like it for pork and chicken highlight the relationships among the graph components, particularly between meat emissions, production, and EI. Total emissions are the product of the GHG efficiency of cattle production (not including feed production) and the amount of meat produced. Meat production in turn is the product of the number of animals slaughtered and the size of the animal in terms of how much meat each one yields. Thus a continent or country that is highly GHG efficient in meat production (i.e., that has a low EI) can still emit a very large amount of total GHGs due to very high production, while a GHG inefficient country (i.e., that has a high EI) can emit very low total GHGs due to having very little beef production. The FAO calculation of GHG emissions from each type of animal from each source is calculated using an emissions factor multiplied by the number of animals of that type. The emissions factors include the effects of general differences in feed quality (which influences how much methane is produced during digestion), and regional climate (which influences the GHG emissions from manure sources).

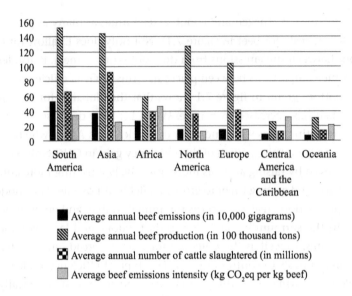

Figure 7.2
Beef emissions, production, and emissions intensities by continent, 2011–2015 averages
Source: FAO 2019.

Figure 7.2 shows that South America produces the largest amount of beef and the most beef emissions, followed by Asia. North America and Europe produce the third and fourth most beef and the fourth and fifth most beef emissions. These top beef-producing continents were not surprisingly home to the top beef-producing countries in 2014: The United States (11.5 MMT[5]), Brazil (9.7 MMT), the European Union (7.4 MMT), and China (6.9 MMT). These four countries also produced the most individual beef emissions, but in a different order from their beef production due to differences in their beef EIs: Brazil produced 316,000 Gg CO_2eq (60 percent of the beef emissions from South America), China produced 161,000 Gg CO_2eq (44 percent of Asia's beef emissions), the United States produced 134,000 Gg CO_2eq (88 percent of North America's beef emissions), and the EU produced 120,000 Gg CO_2eq (77 percent of Europe's beef emissions). Altogether, these four countries produced 42 percent of the world's beef, and 37 percent of the world's beef emissions. Brazil, China, and the United States are also the homes of the three largest meat processing firms in the world: JBS, WH Group, and Tyson (see chapter 2).

Figure 7.3 shows that pork production and emissions are highest in Asia, Europe, and North America. Within these continents, China, the EU, and the United States are the top producers of both pork and emissions. In 2014, China produced 55.4 MMT of pork and 56,000 Gg CO_2eq (74 percent of Asia's pork emissions), the EU produced 22.6 MMT of pork and 37,000 Gg CO_2eq (83 percent of Europe's pork emissions), and the United States produced 10.4 MMT of pork and 21,000 Gg CO_2eq (84 percent of North America's pork emissions). While these three countries have relatively low pork EIs, the sheer scale of their pork production results in them all together producing 75 percent of the world's pork, and 64 percent of the world's GHG emissions from pork. The high pork production and emissions from China clearly shows the result of China's pork boom described in chapter 4.

Figure 7.4 shows that Asia is the top-producing region for both chicken and chicken emissions, followed by South America, North America, and Europe. Similar to beef, the top meat-producing counties in 2014 were the United States (17.7 MMT), China (12.8 MMT), and Brazil (12.5 MMT).

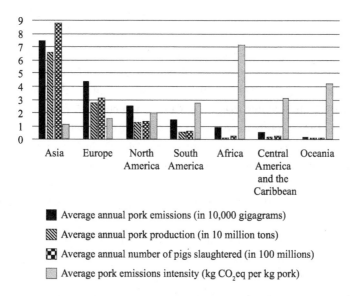

■ Average annual pork emissions (in 10,000 gigagrams)
▧ Average annual pork production (in 10 million tons)
▨ Average annual number of pigs slaughtered (in 100 millions)
▢ Average pork emissions intensity (kg CO_2eq per kg pork)

Figure 7.3
Pork emissions, production, and emissions intensities by continent, 2011–2015 averages
Source: FAO 2019.

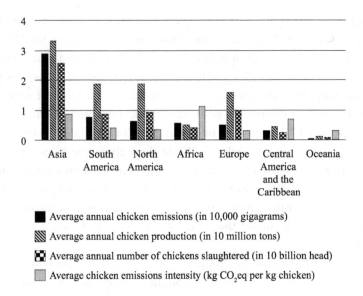

Figure 7.4
Chicken emissions, production, and emissions intensities by continent, 2011–2015 averages
Source: FAO 2019.

However, the countries with the most chicken emissions in 2014 were China (8,000 Gg CO_2eq, which is 24 percent of Asia's chicken emissions), Indonesia (7,000 Gg CO_2eq, which is 21 percent of Asia's chicken emissions), the United States (6,000 Gg CO_2eq, which is 65 percent of North America's chicken emissions), Brazil (4,000 Gg CO_2eq, which is 46 percent of South America's chicken emissions), Iran (4,000 Gg CO_2eq, which is 10 percent of Asia's chicken emissions), and the EU (3,000 Gg CO_2eq, which is 32 percent of Europe's chicken emissions).[6] These six countries produced 58 percent of the world's chicken meat and 56 percent of chicken emissions in 2014.

The high emissions from Indonesia and Iran are the result of relatively high chicken production (they were 10 and 9 respectively in chicken production in 2014) along with high EIs: Indonesia's EI was 3.6 kg CO_2eq per kg of chicken, while Iran's was 1.7. In contrast, China's EI was 0.6, while the United States and Brazil had chicken EIs of 0.3. A similar situation can be seen at the continent level for beef (figure 7.2) where Africa's high beef EI puts it third highest for beef emissions even though it is fifth in beef production. When meat production is very low, even a very high EI can result in low emissions, such as can be easily seen for pork production in Africa, Central America and the Caribbean, and Oceania (figure 7.3).

While the preceding discussion is highly informative, it does not tell us what it is about these regions and countries that make them more or less GHG efficient in producing different kinds of meat. EIs have gone down over time, but why? Is it the result of the rise of high-intensity meat production? In the following section I use a statistical analysis to shed some light on this question and help us to better understand what characteristics are related to high and low emissions intensities.

Statistical Analysis Approach and Design

I combine two approaches for this analysis of what drives the GHG EIs from meat animals. First, I draw on the insights of the political economy of agriculture from the sociology of agriculture literature. Second, I use the STIRPAT model of environmental impact from the environmental sociology literature (Dietz and Rosa 1997; York, Rosa, and Dietz 2003a,b).

The political economy of agriculture literature focuses on the shift toward more intensive production—intensive both in the use of space and in use of capital—and producer and consumer resistance to this shift through alternative food networks and marketing strategies (see, for example, Ilbery and Maye 2005a,b; Little et al. 2012; Watts, Ilbery, and Maye 2005). STIRPAT stands for "stochastic impacts by regression on population, affluence and technology" and represents environmental impacts (I) as being the result of the combined effects of population (P), affluence (A) and technology (T) in a form that can be implemented with a regression analysis (Dietz and Rosa 1997; York, Rosa, and Dietz 2003a,b).

While the sociology of agriculture has long worked from a place of concern over the environmental and social sustainability of modern agriculture (see Hinrichs and Welsh 2003 as a good example), it has tended to give much more explicit attention to the social arrangements of agriculture production and food systems. Combining a political economy view of technology and production practices with the STIRPAT model provides both a conceptual foundation and an empirical approach for the statistical analysis.

In the following statistical model, I use EIs as the outcome variable, in this case representing the unit, rather than the overall, environmental impact (I) of meat production, for each of three types of meat: beef, pork, and chicken.[7] The independent variables of greatest interest in this analysis are animal density, production intensity, manure management, and carcass yield (see tables 7.1 and 7.2 for variable definitions). These first three variables are acting as measures of the extensiveness vs intensiveness of the methods of animal production. The manure management percent variable is the most direct measure available of animal production practices since this approximates the proportion of the meat animal population that is being kept in a confinement situation (i.e., intensive production). High values on all three of these measures indicate more intensive meat animal production, such as confined animal feeding operations (CAFOs) and possibly accompanying higher degrees of corporate concentration (see chapter 2), while lower measures suggest more extensive production practices. Carcass yield is expected to be an important measure when it comes to EIs, because it is also a measure of production efficiency. The rest of the variables are controlling for general national and agricultural context, and climate and culture differences that cannot be included in the analysis directly due to lack of data.[8]

Results and Discussion

The analysis results are shown in table 7.3.[9] The beef analysis is the best in that it explains the most variation in the data at 82 percent (R-squared value), while the pork and chicken analyses explain 44 percent and 80 percent of the variation, respectively. The question we want to answer is: What variables are related to a country having higher or lower meat EIs, all else being equal? Does greater production intensity reduce the EI through more GHG-efficient meat production? Remember that the EI measure does not include emissions from things like feed production or transportation. In considering the results of the analysis, the sign of the coefficient matters: a positive coefficient indicates that as the independent variable increases so does EI, while a negative coefficient indicates that an increase in the input variable is associated with a reduction in EI. The numerical value of the coefficient indicates the relative size of the effect on EI.

Table 7.1
Variable definitions for emissions intensity analyses of for beef, pork, and chicken

Variable name	Beef variable descriptions	Pork variable descriptions	Chicken variable descriptions
Emissions intensity	kg of emissions (converted to CO_2 equivalence) per 100 kg of beef. Logged.	kg of emissions (converted to CO_2 equivalence) per 100 kg of pork. Logged.	kg of emissions (converted to CO_2 equivalence) per 100 kg of chicken meat. Logged.
Animal density	Number of nondairy cattle per 1,000 hectares of agricultural land. Logged.	Number of market swine per 1,000 hectares of agricultural land. Logged.	Number of broiler chickens per 1,000 hectares of agricultural land. Logged.
Production intensity	Nondairy cattle as a percentage of total cattle. Logged.	Omitted from analysis due to lack of variation.	Broiler chickens as a percentage of total chickens. Logged.
Manure management percent	Percent of nondairy cattle emissions that are from manure management. Logged.	Percent of market swine emissions that are from manure management. Logged.	Percent of broiler chicken emissions that are from manure management. Logged.
Carcass yield	Average beef carcass weight in hectograms per animal slaughtered. Logged.	Average pig carcass weight in hectograms per animal slaughtered. Logged.	Average chicken carcass weight in 0.1 grams per animal slaughtered. Logged.
Animal number	Number of nondairy cattle. Logged.	Number of market swine. Logged.	Number of broiler chickens. Logged.
GDP per capita	Gross domestic product per capita in constant 2011 international dollars. Logged.		
Percent agricultural land	Agricultural land area as a percentage of total country land area. Logged.		
Percent arable land	Arable land area as a percentage of total agricultural land area. Logged.		
Island	Is the country an island? 1 = yes, 0 = no		

Table 7.2

Continents, regions, and countries used in analysis (consistent with FAO regions)

Africa	
Eastern Africa	Burundi, Comoros[c], Djibouti[bc], Eritrea[c], Ethiopia, Kenya, Madagascar, Malawi, Mauritius, Mozambique, Rwanda, Seychelles, Uganda, United Republic of Tanzania, Zambia, Zimbabwe
Middle Africa	Cameroon, Central African Republic, Chad, Congo, Democratic Republic of the Congo, Equatorial Guinea, Gabon, Sao Tome and Principe
Northern Africa	Algeria, Egypt, Libya[c], Morocco, Tunisia
Southern Africa	Botswana, Lesotho, Namibia, South Africa, Swaziland
Western Africa	Benin, Burkina Faso, Cabo Verde, Côte d'Ivoire, Gambia, Ghana, Guinea, Guinea-Bissau, Liberia, Mali, Mauritania[c], Niger, Nigeria, Senegal, Sierra Leone, Togo
Americas	
Caribbean	Antigua and Barbuda, Bahamas, Barbados, Cuba, Dominica, Dominican Republic, Grenada, Haiti, Jamaica, Puerto Rico, Saint Kitts and Nevis, Saint Lucia, Saint Vincent and the Grenadines, Trinidad and Tobago
Central America	Belize, Costa Rica, El Salvador, Guatemala, Honduras, Mexico, Nicaragua, Panama
South America	Bolivia (Plurinational State of), Brazil, Chile, Colombia, Ecuador, Guyana, Paraguay, Peru, Suriname, Uruguay, Venezuela (Bolivarian Republic of)
North America	Bermuda, Canada, United States of America
Asia	
Eastern Asia	China (mainland and Taiwan), Japan, Mongolia, Republic of Korea
Central Asia	Kazakhstan, Kyrgyzstan, Tajikistan, Turkmenistan, Uzbekistan
Southeast Asia	Brunei Darussalam, Cambodia, Indonesia, Lao People's Democratic Republic, Malaysia, Philippines, Singapore, Thailand, Timor-Leste, Vietnam
Southern Asia	Afghanistan[c], Bangladesh[c], Bhutan, India, Iran (Islamic Republic of)[c], Nepal, Pakistan[c], Sri Lanka
Western Asia	Armenia, Azerbaijan, Bahrain[c], Cyprus, Georgia, Iraq[c], Israel, Jordan[c], Kuwait[c], Lebanon, Oman[c], Qatar, Saudi Arabia[c], Turkey, United Arab Emirates[c], Yemen

Table 7.2 (continued)

Europe	
Eastern Europe	Belarus, Bulgaria, Czech Republic, Hungary, Poland, Republic of Moldova, Romania, Russian Federation, Slovakia, Ukraine
Northern Europe	Denmark, Estonia, Finland, Iceland, Ireland, Latvia, Lithuania, Norway, Sweden, United Kingdom
Southern Europe	Albania, Bosnia and Herzegovina, Croatia, Greece, Italy, Malta, Montenegro, Portugal, Serbia, Slovenia, Spain, Republic of Macedonia
Western Europe	Austria, Belgium, France, Germany, Luxembourg, Netherlands, Switzerland
Oceania	
Australia and New Zealand	Australia, New Zealand
Melanesia	Fiji, Papua New Guinea, Solomon Islands, Vanuatu
Combined Micronesia and Polynesia	Kiribati[a], Micronesia (Federated States of), Samoa, Tonga

[a]Country not included in beef analysis.
[b]Country not included in chicken analysis.
[c]Country not included in pork analysis.

Animal Production Intensity Measures

Across the three meat EI models, animal density and production intensity are the two consistently significant and positive variables. Higher animal densities and production intensity are both associated with higher EIs. The coefficients are less than 1, indicating that the increase in EI is not proportional to increases in animal density or production intensity. For example, a 1 percent increase in chicken density is associated with a 0.30 percent increase in chicken EI. Beef production intensity has the largest coefficient in the beef model, with a 1 percent increase in beef cattle production intensity associated with a 0.84 percent increase in beef EI. This means that the intensity of cattle production is the main driver of beef EIs and indicates that while more intense cattle production practices do provide some GHG efficiency gains, these gains only slow the rate of increase and are not enough to actually reduce the EI of beef (in which case the coefficient would be negative).

Table 7.3

Prais-Winsten unstandardized regression coefficients and standard errors for beef (n = 3,758), pork (n = 3,417), and chicken (n = 3,759) emissions intensities measured as kilograms of CO_2 equivalent released per 100 kilograms of meat produced (100 kg CO_2eq/kg of meat), 1992–2015

	Beef	Pork	Chicken
Animal density	0.164 (0.021)***	0.134 (0.043)**	0.300 (0.035)***
Production intensity	0.839 (0.048)***	[omitted]	0.751 (0.049)***
Manure management percent	0.165 (0.090)	1.867 (0.910)*	-0.646 (0.379)
Carcass yield	-0.658 (0.046)***	-1.018 (0.082)***	-0.907 (0.05)***
GDP per capita	-0.174 (0.024)***	-0.091 (0.040)*	-0.361 (0.031)***
Percent agricultural land	-0.060 (0.020)**	0.075 (0.036)*	0.077 (0.036)*
Percent arable land	-0.101 (0.026)***	-0.100 (0.049)*	-0.235 (0.029)***
Animal number	0.056 (0.010)***	0.028 (0.019)	0.023 (0.016)
Island	-0.287 (0.057)***	-0.135 (0.076)	-0.517 (0.071)***
Eastern Africa	0.103 (0.102)	0.946 (0.334)**	-0.591 (0.395)
Middle Africa	0.503 (0.113)***	1.436 (0.296)***	-0.082 (0.402)
Northern Africa	-0.445 (0.145)**	1.531 (0.453)**	-0.876 (0.390)*
Southern Africa	0.391 (0.212)	0.706 (0.282)*	-0.781 (0.405)
Western Africa	0.340 (0.096)***	1.032 (0.297)**	-0.407 (0.385)
Caribbean	1.073 (0.145)***	0.867 (0.419)*	-0.122 (0.227)
Central America	0.445 (0.147)**	1.136 (0.443)*	-1.095 (0.206)***
South America	0.403 (0.150)**	0.871 (0.426)*	-0.650 (0.217)**
North America (reference)			
Western Asia	-0.419 (0.115)***	1.101 (0.253)***	-0.822 (0.334)*
Southeast Asia	-0.262 (0.181)	-0.651 (0.142)***	-0.724 (0.211)**
Southern Asia	-0.228 (0.092)*	0.157 (0.185)	-0.406 (0.228)
Eastern Asia	-0.365 (0.149)*	-0.555 (0.358)	-0.091 (0.268)
Central Asia	-0.507 (0.139)***	-0.663 (0.316)*	0.451 (0.320)
Eastern Europe	-0.198 (0.148)	0.031 (0.321)	-0.402 (0.100)***
Western Europe	-0.392 (0.120)**	-0.481 (0.139)**	-0.565 (0.105)***
Northern Europe	-0.105 (0.103)	-0.151 (0.175)	-0.307 (0.093)**
Southern Europe	-0.210 (0.121)	0.024 (0.109)	-0.375 (0.122)**
Australia and New Zealand	0.147 (0.119)	-0.031 (0.130)	0.114 (0.144)
Melanesia	0.678 (0.101)***	0.375 (0.146)*	0.306 (0.141)*

Table 7.3 (continued)

	Beef	Pork	Chicken
Micronesia and Polynesia	0.855 (0.163)***	0.237 (0.222)	0.239 (0.108)*
Constant	4.859 (0.396)***	-0.239 (3.812)	8.957 (1.517)***
Rho	0.904	0.913	0.904
R-squared	0.817	0.436	0.795
Chi-squared	6059.010***	32251.060***	19823.630***

*$p < 0.05$; **$p < 0.01$; ***$p < 0.001$ (two-tailed)

Manure Management

The percentage of animal emissions that come from manure management has a negative effect on the chicken EI and a positive effect on both beef and pork, but it is only statistically significant in the pork model, where it has a large effect compared to the other variables. A 1 percent increase in the manure management percentage from pigs is associated with a 1.8 percent increase in pork EI. This indicates that the more pigs that are kept in conditions that require manure management systems the higher the EI will be, and that there is not an economy of scale gain in EI from keeping more pigs together.

Carcass Yield

Carcass yield is the only livestock-related variable that is related to lower EIs, with negative and significant effects in all three models. The effect is largest for pork with a 1 percent higher carcass weight associated with a 1 percent lower pork EI. Carcass yield is the largest effect in the chicken model, with a 1 percent higher carcass yield associated with a 0.91 percent lower chicken EI. The reduction in beef EI is not as high, with a 1 percent higher beef carcass weight associated with a 0.66 percent lower beef EI.

Conceptually, the effect of carcass yield on EI makes sense because while larger/heavier animals produce slightly more emissions than smaller animals (they typically eat more, thus produce more manure, etc.) the greater amount of meat they produce results in lower emissions per unit of meat. For example, if two large cattle yielded the same amount of meat as three smaller cattle, the emissions per pound of beef would be lower for the meat from the two large animals, since only two metabolic processes are being supported rather than three for the same amount of meat. However, this example assumes that all the animals are the same age when they are slaughtered; obviously, larger animals that were older when they were slaughtered will have lost some or all of that size efficiency in GHG emissions.

Controlling for Agricultural Context
Affluence—measured as GDP per capita—has a reducing effect on EI for all three meats, possibly through the contributions of livestock production research and technology including, but not limited to, genetic improvements related to meat production through increases in carcass yield.[10] The portion of agricultural land in a country has a statistically significant but small negative effect on beef EI, and small, positive effects on pork and chicken EIs. The portion of agricultural land that is arable in a country has a consistently significant and negative effect on the EIs of all three meats. The significance of these agricultural land variables indicates that the biophysical environmental context of a country's agricultural system matters to the GHG efficiency of its meat production, all else being equal, but without more detailed information on the animal production practices within each country the specific mechanism(s) of these effects cannot be determined. For beef, the number of cattle has a small, but significant and positive effect, on EI.

Controlling for Country Location and Other Factors
Island and region dummy variables are used to control for factors related to geographic location, climate, and cultural and economic conditions that are not otherwise included in the model. Negative coefficients of these variables indicate that the region has a lower EI than the reference region (non-islands, and North America), and positive values indicate higher EIs.[11] The significance of these control variables indicates that there are one or more factors specific to these locations that are not otherwise included in the models.[12]

Islands have significantly smaller beef and chicken EIs than non-islands, all else being equal. Because North America has relatively low EIs for all three meats (see figures 7.2, 7.3, and 7.4), it is interesting to see which regions are associated with having even lower EIs—there is something about these places that make them more GHG-efficient meat producers, other than the factors that are already in the model. The Northern Africa, Western

Asia, Southern Asia, Eastern Asia, Central Asia, and Western Europe regions are all associated with having beef EIs that are significantly lower than those of North America. For pork, only the Southeast Asia, Central Asia, and Western Europe regions are associated with having pork EIs that are significantly lower than North America. For chicken, the Northern Africa, Central America, South America, Western Asia, Southeast Asia, and all four European regions have chicken EIs that are significantly lower than North America's.

Conclusions

The world's top meat-producing regions also generate the most GHG emissions from cows, pigs, and chickens: North America (i.e., the United States), South America (i.e., Brazil), Asia (i.e., China), and Europe (i.e., the EU). These are also places that are known for intense meat production and high corporate concentration (see chapters 2 and 4). However, these regions and countries are among the more GHG-efficient meat producers, according to the narrow accounting of GHG emissions from meat used by the FAO, which includes only emissions coming directly from the animals and their manure. What makes them more efficient by this measure?

Technology and intensification are often expected to reduce a number of dimensions of environmental impacts through improving the efficiency of production (Grau, Kuemmerle, and Macchi 2013; Horlings and Marsden 2011). However, for all three meats, the preceding analyses indicate that while some economies of scale in GHG emissions from more intensive livestock production are possible, these efficiencies do not reduce the GHG EIs of beef, pork, or chicken, but only reduce the amount of increase. This reduced increase has certainly not kept up with the increases in global meat production. This can be seen in figure 7.1, where meat EIs have decreased globally over time but total emissions have increased.

Carcass yield was the only production-specific variable that was associated with lower EIs across all three models, which makes sense from a strictly GHG emissions perspective. Carcass weight also represents technology, in that carcass weight can be influenced by animal genetics along with other factors such as diet. Animals that have been bred for meat production are often selected for larger size and muscle mass, and higher feed-to-meat conversion ratios.[13] However, this effect can be confounded, such as when former dairy cattle are slaughtered for meat. These animals typically have a low carcass yield (they have been bred to put their metabolic energy into milk not meat), but their emissions have been spread out over a huge number of milk calories produced along with the meat, making the meat very efficient on a GHG emissions-per-calorie basis.

Importantly, GHG efficiency is only one type of efficiency, just as global climate change is just one type of environmental impact from meat production. Other types of environmental impact from animal agriculture, such as manure runoff into local water sources, poor local air quality, or land-use change are also important, as are issues of animal and worker well-being, and maintaining food sovereignty in the face of greater corporate consolidation and control over the global food supply.

Notably, this analysis has some limitations. First, the EI variable does not include things like feed production, and certainly not things like land-use change done to facilitate animal production (such as converting forests to pastures). Animals raised in the most intensive systems, CAFOs, require lots of calorically concentrated food that has to be grown somewhere else and then brought to them. Not only does the transportation of this feed release GHGs, but so does the production of the feed, especially if it is a crop like corn that is commonly grown with large amounts of nitrogen fertilizer (Robertson et al. 2013).[14] Not only are these feed production emissions not included in the EI measure, but also, when countries import much of their feed, they can appear to have low emissions from fertilizers (see part I introduction; chapter 4).

Second, while permanent grasslands and pastures have the potential to store carbon (i.e., be "carbon sinks") depending on management practices and grazing intensity (Conant and Paustian 2002; Soussana, Tallec, and Blanfort 2010), pastures produced by clearing forest can represent a serious GHG impact, as well as biodiversity impact. For example, Brazil lost 53 million hectares of forest and gained 12 million hectares of pasture between 1990 and 2015 (author's calculations using FAO data). These emissions would also not be counted in the GHG emissions data used in this analysis.

Considering the details and specific context of a production system, both socially and environmentally, is an important complement to the global perspective offered in this chapter. Local effects of meat production and climate change on people and the environment can easily be obscured by national data. For this reason, studies taking a close look at animal production practices and how they intersect with the local social, economic, and environmental contexts are important (see chapter 5).

In conclusion, GHG emissions from meat production are a topic of great importance as meat production continues to rise globally, along with other sources of GHG emissions. While more intensive livestock production may provide some economies of scale in GHG emissions from meat production, it is clear that these efficiencies are not reducing the overall GHG EIs of beef, pork, or chicken, but only reduce the amount of increase. All else being equal, increased carcass yield may reduce the EIs of meat production, but given the high correlations between meat production and total meat emissions the most sure-fire way to reduce GHG emission from meat is to produce less of it.

Notes

1. Ruminants—cows, goats, and sheep, for example—have a multi-chambered stomach in which microbes assist in breaking down food. This enteric fermentation produces methane as a by-product. Animals with a single-chambered stomach, such as pigs, rabbits, and horses, also produce some methane from their digestion, but much less than ruminants. Poultry digestion includes very little enteric fermentation (Hartung 2003), and the FAO does not report enteric fermentation data for poultry.

2. Between 1961 and 2015, enteric fermentation dropped from 63 percent to 60 percent of total meat animal emissions, while emissions from manure applied to pastures increased from 20 to 23 percent.

3. It is important to note that the FAO GHG emissions data used in this chapter does not include GHG emissions from feed production or transportation. High-intensity meat production will thus have emissions from these sources that are not accounted for and may make places with high-intensity meat production look like they are more GHG efficient than they actually are.

4. The decrease in Asia's pork EI has been sustained since the late 1970s, which corresponds with the pork boom in China described in chapter 4. This increase in pork production in China involved the adoption of industrial production techniques, which contributed to lower EI in pork production, even as total GHG emissions from pork increased.

5. MMT stands for million metric tons. One metric ton (or tonne) is equal to 1,000 kilograms or about 1.1 U.S. tons.

6. These six countries stand apart in terms of chicken emissions. The country next below the EU in chicken emissions, India, produced half as many emissions in 2014.

7. The EI variables were calculated by the FAO based on their data on meat production and GHG emissions calculated using the IPCC Tier 1 formulas and emission factors that include climate/temperature and broad characterizations of livestock production practices (IPCC 2006).

8. Based on the political economy of agriculture literature and/or the STIRPAT literature, I include animal population (P), rather than human population, gross domestic product (GDP) per capita as a control measure of affluence (A), and the animal production variables for technology (T). I also control for different agricultural contexts with the agricultural land variables, and use dummy variables for island location and FAO regions to account for climate and culture differences that cannot be included in the analysis directly due to lack of data.

9. This analysis uses a Prais-Winsten regression in Stata14, which accounts for observations within a country being related to each other and thus not truly independent (as is assumed for ordinary least squares regression) (Beck and Katz 1995, 1996, 2004). The data come from *FAOSTAT*, with different samples for each analysis based on the available data. All three samples cover the period from 1992 to 2015. The beef sample includes data for 177 countries (n = 3,758 country-year observations), the pork sample includes 160 countries (n = 3,417 country-year observations), and the chicken sample includes 177 countries (n = 3,759 country-year observations). All countries included in the analysis had at least seven years of data and all regions included complete data for at least two countries. All nonbinary variables were log transformed prior to analysis, in keeping with the STIRPAT tradition, which makes the analysis results be interpreted as "ecological elasticities" (York, Rosa, and Dietz 2003b). The coefficients of the dummy variables are the multiplier of the dependent variable when the dummy variable equals 1, once the antilog of the coefficient ($e^{coefficient}$) is taken (York, Rosa, and Dietz 2003b).

10. GDP and carcass yield are moderately correlated for beef (correlation coefficient = 0.61; GDP: VIF = 2.01 and tolerance = 0.498; beef carcass yield: VIF = 1.73 and tolerance = 0.579), pork (correlation coefficient = 0.59; GDP: VIF = 1.82 and tolerance = 0.550; swine carcass yield: VIF = 1.94 and tolerance = 0.516), and chicken (correlation coefficient = 0.73; GDP: VIF = 2.19 and tolerance = 0.456; chicken carcass yield: VIF = 1.75 and tolerance = 0.570), but not enough to notably influence the results of the analysis.

11. To interpret the size of this difference, take the antilog of the coefficient (see note 9). For example, the antilog of the island variable in the beef model is 0.750 or 75 percent of the beef EIs of non-island countries, all else being equal.

12. This is not a problem, as we are capturing much of this variation with the dummy variables. However, to determine what these factors are would require different models using more detailed data.

13. In the case of chickens, this genetic selection can cause animal welfare concerns when combined with high-density production settings—the rest of the chicken's body does not develop well enough to support their large chests, making walking difficult if not impossible for them.

14. Because it is common for only about 50 percent of the nitrogen applied to a grain crop to be used by the crop, much of the rest is lost to the environment, either to streams and groundwater or to the atmosphere in forms that include nitrous oxide (N_2O), which has a global warming potential that is nearly three hundred times that of CO_2 (Robertson et al. 2013).

8 Livestock Intensification Strategies in Rwanda: Ethical Implications for Animals and a Consideration of Potential Alternatives

Robert M. Chiles and Celize Christy

Over the past several decades, meat production and consumption have come under increasing scrutiny over concerns about public health, sustainability, and ethical obligations toward animals. In the global south, the demand for meat continues to rise alongside growing populations and higher incomes (Nam, Jo, and Lee 2010; Rae 2008), and the industrial model of meat production is assuming increasing prominence in these regions (Fraser 2008; Li 2009).

While much of the social science literature focuses on the pros and cons of traditional, small-scale versus industrial food systems, the animal rights movement historically has regarded both forms of animal agriculture as inherently problematic with respect to ethical obligations toward animals. Supporters of this movement have long premised their arguments on the concept of *speciesism*—the notion that it is just as arbitrary to discriminate against the basic interests and suffering of another being on the basis of species as it is to discriminate on the basis of sex, race, nationality, or creed (Regan 1983; Singer 1975). While nonhuman animals obviously do not have an interest in civil rights like voting, for Singer and others, all sentient animals are equal when it comes to their basic desire to avoid pain and suffering. To quote from Singer (1975, 8),

> If a being suffers there can be no moral justification for refusing to take that suffering into consideration. No matter what the nature of the being, the principle of equality requires that its suffering be counted equally with the like suffering—insofar as rough comparisons can be made—of any other being. ... Nearly all the external signs that lead us to infer pain in other humans can be seen in other species, especially the species most closely related to us—the species of mammals and birds. ... The nervous systems of animals evolved as our own did, and in fact the evolutionary history of human beings and other animals, especially

mammals, did not diverge until the central features of our nervous systems were already in existence.

While there are strong and well-established arguments to be made for reducing animal suffering and death by shifting toward a more plant-based diet, these arguments are usually addressed to consumers in the global north who have access to a diverse array of foodstuffs. It is primarily for this reason that Morris and Kirwan (2006) have described veganism as a largely consumer-oriented movement, and as such, it is often regarded as separate from the sustainable agriculture movement and, more broadly, completely missing from scholarship focused on agricultural development in the global south. To be sure, some advocates of sustainable agriculture have sought to reduce their meat consumption without eliminating it entirely (Bourette 2009; Pollan 2006), and some vegans support organic and local food production, but generally a focus on vegan agriculture has remained separate from much of the mainstream agriculture literature, especially among scholars focused on the global south.

Opponents of plant-based diets level many critiques against the idea of promoting vegan diets, the most common of which is that meat is a necessity in the human diet. By contrast, supporters of plant-based diets argue that vegan diets offer a comprehensive solution that can help to alleviate the practical and ethical challenges associated with meat production, many of which are raised in this volume. Moreover, according to the World Health Organization, the Food and Agriculture Organization of the United Nations, American Dietetic Association, American Heart Association, USDA Dietary Guidelines, American Diabetes Association, Academy of Nutrition and Dietetics, Mayo Clinic, Cleveland Clinic, and Harvard Medical School, vegan diets provide all of the nutrients that are essential for human nutrition and are appropriate for all stages of the life course, including pregnancy. As interest in plant-based diets has grown, so has the quality and quantity of vegetarian and vegan specialty products in many grocery stores and restaurants.

Advocates for plant-based diets are also often accused of elitism and ignoring broader political-economic structures. Scholars have argued, for example, that processed "fake meat" products are expensive, less healthy, support large agrifood corporations, and ironically reify the cultural legitimacy of meat by attempting to mimic it (Morris and Kirwan 2006). Similarly, it has been pointed out that this growth of consumers choosing to be vegetarian has occurred in the context of menu pluralism situated in "affluent, consumer-oriented economy which can draw on a variety of food items, freed by the channels of international trade from the narrow limits of locality, climate and season" (Beardsworth and Keil 1992, 289–290).

Perhaps the most serious charge leveled at veganism is that it is simply not a viable solution for innumerable rural communities across the world, where many peasants and pastoralists literally depend on hunting, fishing, and animal husbandry for their basic survival. This state of affairs appears to put humans' ethical obligations toward animals into direct conflict with competing obligations to food security and cultural rights. It is for this reason that animal rights have largely received limited concern or attention from the development community (Kelly 2016). Indeed, to the extent that animal issues have been prioritized by development professionals, it has largely been out of concern for biodiversity and wildlife tourism as opposed to the rights and dignity of individual animals.

Due to the troubled legacies of colonialism and imperialism, these exchanges take place on an unequal playing field, and many communities in the global south have limited economic choices. Jimmy Smith (Smith 2016 n.p.), the director general for the International Livestock Research Institute (ILRI), has asserted:

> A lot of meat and milk that would remain unproductive in a vegan context is produced on these marginal rangelands. For example, 60 percent of Sub-Saharan Africa is covered by drylands where raising livestock is the main, and often the only, land use option available. ... Above all, livestock are essential to many of the world's poorest people and can't simply be cast aside. In low- and middle-income economies, where livestock account for 40–60 percent of agricultural GDP, farm animals provide livelihoods for almost 1 billion people, many of whom are women. Cows, goats, sheep, pigs and poultry are scarce assets for these people, bringing in regular household income, and can be sold in emergencies to pay for school or medical fees. For those who would otherwise have to subsist largely on cheap grains and tubers—risking malnutrition and stunted children—livestock can provide energy-dense, micronutrient-rich food. Animal-source foods are especially important for pregnant women, babies in their first 1,000 days of life, and young children.

Proponents of animal rights and veganism oftentimes dismiss such arguments as strawmen, namely, by asserting that their awareness and advocacy campaigns are only focused on Western consumers—and *not* on people who depend upon animal products for the subsistence purposes. All too often, the end result of these conversations is that both sides simply talk past each other and then disengage. The question thus remains as to what type of reasonable ethical obligations people have toward animals in communities that are striving toward economic development and greater participation in the global community. We argue that the lack of a serious mutual engagement between both sides on the topic of animal rights and humans' ethical obligations toward animals has resulted in an intellectual vacuum with respect to global meat production and consumption.

The Rwandan Context

In order to situate these abstract concerns and principles in practical context, we chose to focus on the case of Rwanda, a small sub-Saharan African nation that is currently engaging in a process of agricultural intensification while simultaneously investing in a knowledge economy. Rwanda is a country that has a longstanding cultural pride in raising livestock. During the Rwandan civil war (1990–1994), cattle were scarce and meat consumption was low, but over the past several decades livestock production has emerged as a key sector in Rwanda's economic portfolio, specifically the poultry sector (MINAGRI 2012).

Despite its many social and economic achievements in the years following the 1994 genocide, Rwanda remains one of the world's poorest countries. It is stricken by widespread childhood malnutrition, and animal protein can provide many of the diversified nutrients that are currently lacking in the diets of many rural Rwandans. In recent years, Rwanda's government has sought to capitalize on the increasing regional demand for meat products across Eastern Africa by intensifying and industrializing livestock production (Mbuza et al. 2017; Thornton 2010). While many of Rwanda's agricultural, demographic, and economic challenges are widespread across the global south, the combination of Rwanda's small geographic size, growing population, and transitioning economy makes it an exemplary case study from which to study the ethics of animal agriculture in the developing world.

Studying the Rwandan livestock industry from the standpoint of development ethics further contextualizes this volume's broader investigation of food sovereignty, governance, and social and environmental justice. While the Rwandan government's push to privatize and sustainably intensify its poultry sector could potentially improve the quality and diversity of people's diets, transitioning away from traditional modes of production could also serve to consolidate ownership and jeopardize local food sovereignty. Moreover, there is no guarantee that conventionally sourced poultry products will be available, accessible, or affordable to those who need them the most. These government initiatives could also exacerbate ongoing competition over Rwanda's limited and marginal lands, both with respect to ownership as well as usage. There are also no guarantees that the treatment of animals will improve with sustainable intensification, and if this program is successful, it will by definition result in more animals being raised and killed.

In what follows, we situate the Rwandan experience in the socioeconomic, ecological, and historical context of development. Rwanda is similar to other sub-Saharan African countries in that its pathway to development is occurring amid sweeping changes: rapid population growth, urbanization, global environmental crises, and a nutrition transition. Next, we discuss how Rwanda has prioritized the poultry industry as a pathway toward food security and economic growth. We then consider the ethical questions that this type of investment raises with respect to humanity's ethical obligations toward animals. We conclude the chapter by reflecting on the past, current, and future potential of non-livestock-oriented approaches to Rwandan food security and community-economic development. Our overarching purpose with this chapter is to problematize the assumption that intensifying livestock production is the only viable future for the citizens of Rwanda and the global south more broadly.

The Broader Context of Livestock-Oriented Development Strategies in Rwanda

The demand for meat and other animal-source proteins is surging across the global south, a phenomenon that Weis (2015) describes as the "meatification" of the global diet. Indeed, economic and cultural globalization is paving the way for people in the global south to increase both the total demand for food products and the *composition* of this demand. Here, more and more consumers around the globe are shifting away from traditional starches toward resource-intensive animal products, crop cereals, oils, fresh fruit, vegetables, and convenient processed foods (Bett 2012; OECD/FAO

2016; Thornton 2010). In a meta-analysis of 393 different studies on meat consumption and income elasticity, Gallet (2010) found that larger incomes resulted in particularly higher levels of spending on beef and fish (with less spending on lamb, pork, and poultry).

Smil (2001) breaks these global dietary trends down into two distinctive stages. In the first stage, which is the "expansion" effect, the primary change is that of increased energy supplies, and extra calories come from cheaper foodstuffs of vegetable origin. The second stage is the "substitution" effect, which is caused by a shift in the consumption of foodstuffs with no major change in overall energy supplies. Here, regions will shift from diets primarily comprised of carbohydrate-rich staples to vegetable oils, animal proteins, and sugar. Currently, a majority of sub-Saharan Africa is experiencing the substitution effect, and this is due to affluent consumers' cultural preferences for higher caloric products like meat (Smil 2001). This phenomenon further magnifies the impact of international development programs and initiatives that promote livestock production and animal-protein consumption.

A key driver of the increasing demand for meat commodities is rapid population growth, accelerated urbanization, and increasing per capita income in countries in the global south, and this is particularly evident in sub-Saharan Africa (OECD/FAO 2016). Drawing on cross-national data from the United Nations, the FAO, and the World Bank, York and Gossard (2004) found that countries with highly urbanized populations consume more meat than countries with less urbanized populations, countries in temperate regions consume more meat than nations in arctic and tropical regions, countries with more land area likewise consume more meat, and economic development increases both meat consumption and fish consumption (where Western countries consume meat and Asian nations consume more fish). The nutrition transition may well progress at a greater speed in regions like sub-Saharan Africa, where the consequences of dietary change could also be more impactful as compared with other regions (Popkin 2002). Southern and Western Africa have the largest economies in the sub-Saharan Africa region, and their per capita caloric intake is higher than that of Central and Eastern Africa (OECD/FAO 2016). In Eastern Africa, the per capita caloric intake is projected to expand to almost 7.5 percent (162 kcal/day/person) by 2025 (OECD/FAO 2016).

In many ways, the nutritional transition has been both a blessing and a curse. On the one hand, the consumption of more meat products can benefit impoverished people in the global south who suffer from nutritional problems like anemia, stunting, and wasting. Meat is a valuable source of high-quality proteins, fats, and minerals, like iron, zinc, and all B-vitamins (other than folic acid) that at times aren't attained at adequate levels with traditional diets. On the other hand, in many sub-Saharan Africa countries like South Africa, Ghana, Kenya, or Nigeria, integration into globalized markets has also resulted in increased rates of noncommunicable diseases of "over-nutrition" like diabetes and obesity. Satisfying increasing and changing demands for animal food products while also sustaining the natural resource base (soil, water, air, and biodiversity) is another major challenge facing global agricultural producers (Alexandratos and Bruinsma 2012).

Increasing meat demand is pressuring agricultural producers to modernize and industrialize, and East African countries like Rwanda are currently implementing policies intended to expand and enhance livestock productivity. Many of these policies are being taken on and implemented by various Western-based development organizations like the International Livestock Research Institute (ILRI) and Heifer International. These organizations' activities demonstrate that the expansion of global meat production and consumption is not a "natural and inevitable outcome of development" as per local traditions and cultural preferences; rather, it involves active campaigns of continued Western intervention in the global south (Hansen 2018, 57).

For livestock-oriented development programs, meat, milk, and egg production is the core pillar of food security, financial development, and social stability in the global south. These organizations thus implement livestock programming and research focused on sustainability and environmental health, economic profitability, and socioeconomic equity. ILRI, for example, is committed to food security, market participation, and poverty reduction through the efficient, safe, and sustainable use of livestock. Per its namesake, ILRI concentrates its efforts on agricultural productivity and policy research, which they argue "is helping farmers exploit the potential of their animals to turn the nutrient cycling on their farms faster and more efficiently" (ILRI 2018, 1).

Another large livestock development program is Heifer International. Heifer International's core mission is more applied, and involves gifting livestock to needy families. Recipient families are then obligated to share their knowledge, skills, and animals' offspring with others (Dierolf et al. 2002). Through this practice, Heifer International uses livestock production as a method for capacity building, resilience, community-based planning, food security, and poverty alleviation. Indeed, livestock can provide a household with vital micronutrients while allowing for individuals to participate in local markets by selling goods and products.

Of particular interest to us is the lack of attention and priority these livestock programs give to animal welfare and thinking about human obligations toward animals more generally. We are not alone in expressing this concern. GiveWell (2018, 1), a nonprofit charity evaluator, has raised flags regarding the issues of livestock health and potential underproduction, and verification of recipients' knowledge and commitment regarding animal welfare, among other issues. It may well be that simple cash transfers are a more effective way to eradicate poverty than making livestock donations (see GiveDirectly 2018). In the following section, we consider the social and ethical implications of encouraging the expansion of livestock production in the global south by examining the case of Rwanda in greater detail.

Rwanda: A Case Study in Livestock Intensification

Rwanda is a mountainous landlocked country located in East Africa with the highest population density in Sub-Saharan Africa. Agriculture contributes 81 percent of the country's total GDP, with most of the economy based on subsistence local farming. Despite its many successes over the past several decades, in many ways, the country is still striving to recover from the social and economic trauma of the 1994 genocide. An estimated 40 percent of Rwanda's total population lives below the poverty line (CIA 2016), and this segment of the country is almost entirely dependent on agriculture as a primary source of income and livelihood stability.

Rwanda is also changing rapidly, and all the global challenges that we identified in the previous section (population increase, rising per capita income, urbanization, agricultural intensification, climate change, etc.) are present and occurring. There is intense economic and political competition for what limited land is available, and this competition also increases prices for food-feed crops (Rosegrant et al. 2009). These dynamics are taking place at a much slower pace in other East African countries like Kenya and

Tanzania, both of which already have an established regional and international market, a large number of consumers purchasing high-value products (e.g., meat), and extensive urbanization.

Despite Rwanda's fertile ecosystem, food production often does not keep pace with population growth, and Rwanda currently depends on neighboring countries for animal protein and other food sources (CIA 2016). With minimal land available for grazing, the potential options for expanding the livestock industry are limited. Moreover, while land productivity has increased with both crop and livestock intensification, Rwanda's use of marginal plots and traditional pastoral lands for agriculture has resulted in high rates of soil degradation, erosion, and deforestation (Van Hoyweghen 1999). The poultry industry's minimal land requirements make it a priority for investment by the government (Mbuza et al. 2017).

The United Nations and the International Monetary Fund have identified livestock development as the key pillar of their poverty reduction strategy in Rwanda. The Rwandan government shares this outlook and developed policies and strategies in 2012 to enhance the nation's livestock industry—particularly with respect to the poultry sector (Mbuza et al. 2017). Not only does poultry have a lower price, feed requirement, and climate footprint per kg produced as compared to beef and pork, but the poultry industry as a whole has also been more successful in adopting cost-lowering technologies. Rwanda's agricultural policies have accordingly been drafted with the intention of diversifying its meat industry through the poultry sector, increasing meat production, modernizing the livestock industry's infrastructure, and improving access to domestic and foreign markets. English, McSharry, and Ggombe (2016, 28) agree: "Livestock products (including hides and skins, dairy products, meat and live animals) are among the top emerging non-traditional exports in Rwanda. ... Development of this subsector has large potential to improve household incomes since 65 percent of households in Rwanda are engaged in rearing some type of livestock."

The Rwandan Ministry of Agriculture and Animal Resources (MINAGRI) is thus encouraging and funding Rwandans to expand and intensify the poultry sector of Rwanda's agriculture in order to compete with regional markets and expand the accessibility of animal products (MINAGRI 2012). In a 2012 report, MINAGRI stated that its vision for Rwanda was centered on the following objectives: (1) ensuring meat security for Rwandans, (2) using the livestock sector to combat malnutrition and poverty, (3) developing

Rwandan livestock competitiveness in Africa, (4) promoting the development of a strong and sustainable meat industry, and (5) developing foreign exchanges. The intensification strategy is already well underway, as the commercial/industrial poultry industry is growing seven times faster than smallholder livestock systems (MINAGRI 2012).

A key element of the Rwandan strategy to industrialize and commercialize its livestock production has been a comprehensive program of privatization and liberalization (CAADP 2013). By the same token, Rwanda's agricultural development strategy, investments, and policies have essentially disregarded the village farmers who make up the majority of farmers in the poultry sector. The role and value of smallholder farming systems continues to be neglected, overlooked, and extended little political significance or scientific prestige (Booth and Golooba-Mutebi 2014; Guèye 2000). Moreover, while livestock intensification features prominently in the Rwandan government's development strategy, the nation's political leadership is also seeking to diversify and modernize Rwanda's overall economy. According to the Rwanda Development Board (2018), while agriculture is projected to grow from 5.8 percent to 8.5 percent by 2018, the number of people earning a living primarily by agriculture is expected to decline from 34 percent to 25 percent. Among those who remain in the agricultural sector, there will be fewer farmers and more employment in agro-processing. Exports are expected to increase on average from 19.2 percent to 28 percent per annum; and imports are expected to remain at an average rate of 17 percent growth. The Rwandan government has sought to accelerate the transition away from low-income subsistence agriculture, and its objective over the past decade has been to evolve into a middle-income, knowledge-intensive, service-sector-oriented economy by the year 2020 (MINAGRI 2012).

Rwandan Agriculture in Transition: Implications for Animals

The intensification of Rwanda's livestock sector raises important questions regarding ethical obligations to animals. When it comes to animal welfare in sub-Saharan Africa, many of the policies and laws that are in place largely concern wildlife (e.g. poaching, ivory hunting, and capture) as opposed to domesticated livestock. Masiga and Munyua (2005, 579) argue that "there is an urgent need for African countries to develop, implement, and enforce transport and pre-slaughter handling procedures and to improve handling facilities." Further they argue that "African countries need to develop and implement policies and legal frameworks that address animal welfare issues and, at the same time, encourage compliance through community education and awareness about animal welfare" (ibid). On the one hand, Botswana, Lesotho, Malawi, Namibia, Seychelles, South Africa, Tanzania, Zambia, and Zimbabwe currently have animal protection or animal health acts, while South Africa and Zimbabwe have animal welfare codes of practice, although a World Organization for Animal Health (OIE) report argues that with the exception of Tanzania, most countries' existing policies are outdated (Devereux 2014; OIE 2011). Rwanda, on the other hand, has not yet developed any form of animal welfare codes, acts, policies, or laws.

The lack of animal welfare standards and laws for Rwandan livestock is emblematic of the overall structure of this country's agricultural sector. Despite the rapid growth of the industrial livestock model, Rwanda continues to be dominated by smallholder farms, which are largely disconnected from extension services and training. Most Rwandan farmers either sell live birds and eggs directly to consumers at local markets or to village collectors who act as wholesale distributors (MINAGI 2012). The processing of the poultry meat occurs mostly on the farm level, where it is then distributed directly to retail outlets, or at the household level after live birds are bought in markets. Within the current system of primarily household slaughter and processing, there are no guidelines or regulations associated with animal welfare and handling and there is no training or education system in place to communicate better animal handling. In other words, with no developed/professional entities or companies in processing, packaging, or preservation there are no animal welfare guidelines or methods in Rwanda. This also means there are no training procedures or modern facilities to promote adherence to worker safety, biosecurity, or animal welfare protocols.

The key question to consider is whether or not Rwanda can sustainably intensify its livestock sector—while still addressing global concerns about the ethical treatment of animals—by adopting international animal welfare practices and protocols. For Busch (2008), modernization and the transition from subsistence to market-oriented agricultural production invariably revolves around the question of *standards*, and international expectations for appropriate animal welfare practices is no exception (Ransom 2007). Standardization essentially involves farmers being told by processors and retailers that they need to conform to the latter's standards of production in order to bring their products into the global market. Optimists believe that this can ensure that animal welfare can be standardized and thus improved.

The basis upon which standards are set and determined arguably is a matter of politics and economic power as much as science. Appealing to animal welfare science as the ultimate arbiter of appropriate production standards glosses over the processes by which scientific findings are socially constructed, as this construction is always the result of political struggle between competing interest groups and ideological paradigms (Lassen, Sandøe, and Forkman 2006). As observed by Lassen, Sandøe, and Forkman (2006, 223), "It is now widely recognized that assessments of animal welfare are based on a number of assumptions which are of an ethical nature." The authors continue, "It matters a great deal how animal welfare is defined—whether it is defined in terms of animal function, of the balance of enjoyment of pleasure and suffering or pain, of preference satisfaction, or of natural living" (ibid). For example, "the U.S. government uses health indicators (e.g., presence of illness) to measure animal well-being, whereas the EU government relies upon health, productivity, physiology, and ethology" (Ransom 2007, 34). Another significant debate in the field of animal welfare measurement and assessment concerns whether or not welfare is measured in terms of the average experience of all animals or in terms of those animals that are worst off.[1] Assessments based upon the average experience are biased toward commercial interests (Lassen, Sandøe, and Forkman 2006), as production is organized around management and productivity in the aggregate and not the lives of individual animals. Tolerating a certain level of incorrect practices or outcomes thus results in what Perrow (1984) refers to as "normal accidents." For example, the introduction to part III of this volume notes that up to a thousand sheep of a shipment dying while in transport from Australia to the Middle East is considered a "normal" industry standard.

Standards have been increasingly relied upon in an effort to answer international criticisms of meat production and consumption as concerns are raised about environmental sustainability, animal cruelty, food safety, and social justice. These concerns are particularly acute with respect to the poultry industry. Due to chickens' small size, the mass production of chicken meat and eggs to meet consumer demand requires near-astronomical numbers of animals. It also bears worthy of mention that no standard of production can definitely address the philosophical questions regarding the killing of animals for food in the first instance.

One of the authors of this chapter personally witnessed the animal welfare challenges facing the Rwandan commercial livestock industry. In collaborating with a local Rwandan feed mill on a development initiative, she observed first-hand the project planning and implementation. The focus of this development project was to improve the socioeconomic status of nonfarmer rural residents by training them to raise broiler chickens. This particular project is one of many in Rwanda dedicated to improving social, monetary, and educational capital for people in rural areas. Through their participation in the program, prospective farmers received a loan to fund establishment of their coop and materials for brooding, feeders, and drinkers. As a bystander to the project, the author saw how the contours of the initiative were directly shaped by the funders: a private-public partnership between a U.S. four-year public university and a U.S. international aid development agency and a local Rwandan feed mill. The funders' goals dictated how the farmers were trained, selected, and compensated. While the donors, contributors, and program managers certainly were doing what they thought was best for the farmers, there was no clear institutional process dedicated to integrating farmers' perspectives or concerns. Moreover, while the training reviewed proper poultry management, poultry nutrition, and disease control, there was little to no focus on animal welfare issues like bird handling or stocking density. This personal anecdote serves to highlight many of the concerns this chapter raises regarding ethical obligations toward animals, namely, whether or not sustainable intensification can deliver on its promise of a higher standard of living for people and animals alike.

The moral controversy surrounding the treatment of animals in countries in the global south is only exacerbated by the strains and tensions that come with the bitter and painful legacy of cultural and economic imperialism. In practical terms, we would argue that the key question is *not* whether Western values should be *forcibly* imposed on countries in the global south. All people have a basic human right to autonomy and self-determination, and history shows that imposing external values by force almost inevitably creates far more problems than it ever solves. Rather, the key question for us to consider is what the basic terms of international engagement and investment in agricultural development should be based upon. If we reject cultural relativism as a justification for any and all local traditions and cultural preferences (e.g., denying girls access to education because they are less valued within a community), and we accept the argument that people have ethical obligations toward animals, it follows in turn that appealing to cultural relativism and local tradition is not a viable justification for any and all forms of animal treatment. As such, we would argue that a firm commitment to the moral standing and inherent dignity of animals should be reflected in the policy and practices of international development organizations.

Is There an Alternative? Non-Livestock-Oriented Approaches to Rwandan Food Security and Community-Economic Development

In contrast to countries like Rwanda, the logistical challenges faced in the global north with respect to adopting healthy plant-based diets are quite minimal. Most consumers in the global north have readily available access to a wide variety of vegetables, fruits, whole grains, nuts, seeds, and legumes that can meet all of their nutritional needs. Consumers in the global north are also gaining increasing access to vegetarian and vegan offerings that are ready to "heat and eat." In sub-Saharan Africa countries like Rwanda, plant-based diets are already being consumed, but these diets are largely comprised of starchy vegetables (e.g., sweet potatoes and cassava) and beans. It is cheaper for a household to purchase or grow traditional and local starchy plants, legumes, and pulses than meat products. When consumed on a regular basis, there is not much nutritional or flavorful variety in these crops, and this is where much of the desire to consume meat stems from. Animal-source protein products provide both rural and urban communities in sub-Saharan Africa with enhanced dietary variety that appeals to people's cultural taste buds and increases the availability of nutrients that are not always received in the traditional starch diet. Increase in demand is one reason among many as to why the Rwandan government has positioned itself to become more intensified agriculturally through livestock.

Is it possible for non-livestock-oriented investments to effectively address Rwanda's need for food security and community-economic development and to satisfy consumers' palates? The traditional diet of Rwandans is starch plant-based, and fortifying staple starches can benefit diets without incorporating animal protein products. Food fortification is an innovation of food science and technology that has been used diligently within the past century among countries in both the global north and south to help address nutrient deficiencies. The process of food fortification involves increasing the content of a micronutrient in food in order to improve the nutritional quality of the food supply and provide public health benefits (WHO 2015). Traditionally, agricultural scientists have sought to fortify foods that are already dietary staples in nutrient-deficient countries, and the fortification itself primarily has encompassed nutrients like iron, niacin, and vitamins A, D, and B. The fortification of staple foods with essential vitamins and minerals is a highly cost-effective solution to introduce diverse nutrients into the diet, especially in rural areas, where improving dietary quality through food variety is not always practical or feasible (Andersson et al. 2017; WHO 2015). Moreover, by increasing the diversity of their diets and micronutrient intake, rural villagers can increase their disposable income by being more productive in work while spending less money on medical treatments (Demment, Young, and Sensenig 2003). Rwanda has experienced particular success with the adoption and dissemination of "iron beans," a conventionally bred variety of iron-biofortified beans. Mulambu et al. (2017) found the following:

> Six years after release and thirteen years after initial research activities began, it is estimated that more than 800,000 Rwandan farm households are growing and consuming iron beans, which contain significantly higher amounts of iron than their conventional counterparts. ... An efficacy study showed that women between the ages of 18 and 27 who consumed biofortified beans exhibited increases in hemoglobin and total body iron levels. ... Strong support from the Government of Rwanda to improve nutrition and health has led to rapid integration of biofortification into its agriculture and health programs, complementing existing supplementation efforts.

With respect to community-economic development, Rwanda has already taken great strides towards diversifying its agricultural output and its economy as a whole beyond the livestock sector. Coffee and tea are among Rwanda's top exports, second only to tourism. They also "involve the most people in Rwanda, and probably the greatest number of poor people" (English, McSharry, and Ggombe 2016, 26). With respect to cropping and horticulture, Booth and Golooba-Mutebi (2014, s180) state:

> There is much room for improvement in the production of staple root crops, bananas and grains, [and] commercial horticulture is also promising as a contribution to this effort. ... Rwanda's climate and topography are well suited to

production of a range of fruits, vegetables and flowers. A broad band of cool and humid terrain in the west is suited to European-style fruits and vegetables, including beans, peas, cauliflower, mushrooms, citrus fruit and strawberries. The warm and humid central-south is ideal for tropical fruits such as banana, passion fruit and pineapple. The warm and dry north-east is suited to groundnut, sunflower and pulses.

Such a diverse array of potential agricultural products offers the possibility of a more varied diet, both for nutritional needs and for consumer palates.

Other valuable investments in local food security could be made in smallholder agroecological cropping. Isaacs et al. (2016, 491) observe that "in Rwanda, farmers' traditional farming systems based on intercropping and varietal mixtures are designed to meet a variety of livelihood objectives and withstand risks associated with fluctuation in market and agro-climatic conditions." The authors note, however, that mixed farming systems have been disappearing since 2008 when the Rwandan government mandated intensification strategies. From their own research, the authors found that improved intercropping systems tend to outperform the government-mandated system of alternating sole-cropped bean and maize season by season. This leads Isaacs et al. (2016, 491) to conclude that while Rwanda's agricultural intensification strategy "aims to improve rural livelihoods through agricultural modernization, it fails to acknowledge the multiple and currently non-replaceable benefits that diverse cropping systems provide, particularly food security and risk management."

While agricultural growth and productivity provided a needed boost to Rwanda's post-1994 poverty reduction efforts, the emergence of the Rwandan service industry "became a leading sector in growth for much of the last two decades" (Ggombe and Newfarmer 2017, 10), and continued investment in this sector is essential to its regional economic competitiveness (Ggombe and Newfarmer 2017). As noted by English, McSharry, and Ggombe (2016, 24), the tourism sector "is relatively labour intensive and requires a wide range of skills. Many of the jobs are low skill, but typically better paid than agriculture, thereby contributing directly to poverty reduction." Information and communication technology (ICT) is another rapidly ascending industry in Rwanda, and Rwandan "technological innovation could stimulate and sustain economic diversity and trade creation" (Murenzi and Hughes 2006, 258). One intriguing pathway toward further expansion in the Rwandan ICT sector is tech/innovation hubs, which are "organizations that support entrepreneurs as they develop and launch their ventures" (Obeysekare, Mehta, and Maitland 2017, 1). In listening to Rwandan tech entrepreneurs who participated in these hubs, Obeysekare and colleagues (2017, 1) found that "such communities can promote creating new humanitarian technologies that solve local problems," namely, through cultivating innovation ecosystems, creating new businesses and ideas, providing needed infrastructure, facilitating community and networking, and giving local entrepreneurs a sense of status, prestige, and success.

Rwanda has already begun to transition away from farming as its economic mainstay, and has achieved rapid growth in ICT, but it still has a long way to go before it can attain middle-income status and compete in the global market for ICT and other professional services (English, McSharry, and Ggombe 2016). Further investments in tourism, agro-processing, infrastructure, electrification, financial services, transportation, telecommunications (especially mobile services and broadband), urban services, public-private partnerships, technical and vocational education, and higher education offer particular promise (English, McSharry, and Ggombe 2016; Ggombe and Newfarmer 2017; Murenzi and Hughes, 2006).

Discussion

Sustainable global development requires all-inclusive thinking about conservation, human responsibility, environmental justice, and related challenges that transcends specific issues, interests, and causes (Ziser and Sze 2007). Countries in the global north must therefore be reflective on their moral obligations with respect to the legacy of Western colonialism and imperialism, environmental and climate justice, food sovereignty and cultural sovereignty, and ethical obligations toward animals. The moral question that served as the inspiration for this chapter—namely, whether or not the international development community should emphasize livestock versus alternative approaches to agriculture development—still remains.

Our goal with this chapter was not to answer this question definitely for readers, but rather to help broaden the conversation and provide resources through which students, scholars, practitioners, and other professionals can engage in a more informed exploration of this topic. While there is no silver bullet solution for the constantly evolving questions of humans' obligations toward animals, and Rwanda is no exception, a loose-knit consensus may be coalescing around a shared desire for expanding and democratizing access to education, public health care, fortified foods, and agricultural knowledge and technology. Investing in girls' education and public health care programs, particularly voluntary family planning initiatives, can help to mitigate population growth, improve food security, reduce malnutrition, and increase lifetime wages (PRB 2010). Expanding community knowledge about sustainable agricultural practices—including traditional and indigenous methods of farming—can also provide people with more options and agency over their livelihoods. Lastly, technology transparency is a means by which the global north can share technology that is adaptable and useful for all farmers. Many smallholder farmers in sub-Saharan Africa nations lack access to training, technical knowledge, new cultivars, and other technologies that could help them to improve their farming operations. Exposure to technological training for sub-Saharan Africa farmers can help to mitigate the social and environmental injustices that are faced by both people and animals.

To be sure, many of the agricultural techniques and practices that are currently circulating across the globe, like confined animal feeding operations and genetically modified crops, are controversial. These disagreements are unlikely to be resolved in the immediate future, but accepting and disseminating the intensification of livestock production (and all the negative consequences that come from this intensification) without considering and then investing in viable alternatives seems highly problematic. What governments, companies, agricultural scientists, development organizations, and farmers can do, however, is to recognize and take seriously the fact that all species strive to avoid pain and suffering. Indeed, recognizing the inherent moral worth of human and nonhuman animals is not mutually exclusive. This recognition is but a launching point for ever more inclusive, creative, innovative, and dynamic approaches to equitable and sustainable food systems.

Notes

1. For an alternative conception of animal welfare, which is based upon individual care, see Buller and Morris 2003.

9 Conclusions about the Global Meat Industry: Consequences and Solutions

Elizabeth Ransom and Bill Winders

As noted in the introduction to part III, this book has set about providing a broader understanding of the operation of the global meat industry in different geographical spaces and at different scales. By providing empirical evidence through case studies and statistical data, we have a clearer understanding of the growth of meat global production, which encourages our questioning the central role of meat in our food system in the twenty-first century and the types of policies and practices that will need to be established to create a different type of food system. Specifically, this work has drawn attention to three themes that are present within the operation of the food system today: (1) the role of state and corporate entities in contributing to the growth of global meat; (2) the ways global meat production contributes to reduced food security; and (3) the social and environmental consequences of the global meat industry. This chapter will first revisit these three themes, focusing on the ways the chapters in this book speak to these themes. We then turn our attention to possible solutions to tackling some of the complex problems associated with the global meat industry, first focusing on consumer movements, then shifting to broader movements that do not originate from consumers. Finally, we conclude by discussing the role that climate change may play in inducing change to our global food system.

State and Corporate Involvement the Global Meat Industry

Several chapter authors engage with the role of the state and corporations, one of three major themes in the book. The authors reveal the ways in which state policies and capitalist markets have encouraged overproduction and, in some spaces, overconsumption of aquatic and terrestrial meat. Howard's chapter on "Corporate Concentration in Global Meat Processing" powerfully demonstrates not only the growing concentration of corporate ownership in the meat sector, but also the important role that direct and indirect government support has played in supporting corporate concentration. In doing so, Howard argues that governments increasingly are faced with a legitimation crisis, as continued subsidy support for industrial meat production is coming under scrutiny by citizens due to the biophysical and social limits that global meat firms are rubbing up against.

Moving from corporate ownership of terrestrial species, Bailey and Tran turn our attention to the role of corporations in the future of aquaculture in their chapter "Aquatic CAFOs." The authors reveal how an increase in international trade in aquaculture has led to an increase in corporate investment, both in the CAFOs and in ownership of the genetic material. For example, they note, a German corporation has recently acquired a Norwegian firm that controls 35 percent of the global salmon market. At present, aquaculture is not as concentrated as other meat industries, but Bailey and Tran argue that the conditions are ripe for aquaculture to follow a very similar path towards concentration that you see in poultry production, which Howard details in his chapter.

Schneider's "China's Global Meat Industry" (chapter 4) continues the theme of the role of the state and corporations, with a detailed accounting of how the Chinese state works to support CAFO pork production. As Schneider (79) explains, "While governments always play a role in corporate power ... in China these relationships are not hidden. ... State-corporate relations are central to the political economy of China's reform-era pork boom."

Food Security

The second dominant theme in this book is the declining role of food security in the midst of increasing meat production. While it may seem counterintuitive to argue that increasing production decreases food security, several authors in this work reveal the ways in which this counterintuitive process may work. Schneider's chapter 4 reveals that transformation of the Chinese countryside is due to the growth of CAFOs. While pigs have always played an important role in Chinese culture, only recently has pork meat become a mainstay of local diets through the growth of state-supported CAFOs. However, this industrialization of pork production has profound consequences for smallholder agriculture in rural China, with the consequence being a decline of smallholder production, thereby making smallholders livelihoods' precarious. Clearly, industrial meat production is not alone in contributing to the decline of smallholder production, as this is true of most industrialized agricultural systems that depend on commodity-focused agriculture (Carolan 2012; McMichael 2009a). However, chapters 3, 4, 5, and 8 (by Bailey and Tran; Schneider; Rudel; Chiles and Christy, respectively) in this volume identify the unique and important role that livestock and aquatic species have played in indigenous and smallholder communities. Unlike cotton or soybean fields, smallholders around the world have relied upon livestock and aquatic animals to maintain their existence. Livestock are important sources of energy, fertilizer, food, and income for smallholders. In the case of aquaculture, the small fish used for feed in intensive production reduce the small fish that are available for consumption by poorer communities who have historically relied on these species for food. In terms of fertilizers, livestock are integral to the natural ecosystem of agriculture for smallholders, who generally do not and cannot rely on chemical fertilizers.

Two other dimensions to declining food security include the labor strategies most slaughterhouses around the world deploy in order to ensure a profit: one is low-wage positions; the other is the increase in highly processed meats at cheaper prices. As Freshour explains in "Cheap Meat and Cheap Work in the U.S. Poultry Industry" (chapter 6), worker pay has declined over time, which ironically contributes to household food insecurity due to insufficient money among the households of the slaughterhouse workers. Therefore, slaughterhouse workers (and other low-wage workers) around the world are often dependent upon purchasing cheaper foodstuffs. These cheaper foods include highly processed meats, which may fill a person's stomach, but do not offer the appropriate nutrition. This has led to the rise of the so-called obesity-poverty paradox, mentioned in the introduction to part II, whereby poorer people in industrialized countries, and increasingly in many countries in the global south (e.g. Brazil, South Africa, and India) are more likely to be overweight and suffer from chronic diseases—such as Type II diabetes—than other income groups.

Social and Environmental Consequences

Finally, several chapters speak to the third theme of the book, the social and environmental consequences of industrial meat production as it relates to people, land/territory, and animals. Industrial meat production contributes to a silencing of the tragedy that unfolds daily for humans, animals, and the ecosystems that sustain industrial meat production (Pachirat 2013; Simon 2017). To continue to produce inexpensive meat, the system relies on silencing the voices and suffering of human laborers, externalizing the costs to the environment and the communities within which these production chains are situated, and hiding the toll the system demands of sentient animals' lives (Schlosser 2001; Simon 2017). Unlike the smallholders who rely on animal waste for fertilizer in their agricultural ecosystem, industrial meat production produces animal waste on such a scale that it must be collected into massive holding ponds, threatening to contaminate waterways, agricultural crops (e.g., contamination of vegetables with Escherichia coli [E. coli]), and communities (i.e., excessive odor).

Rudel's chapter 5 and Freshour's chapter 6 reveal the ways in which social inequalities can be exacerbated by the operation of the global meat industry. Rudel's chapter "Amerindians, Mestizos, and Cows in the Ecuadorian Amazon" brings racial-ethnic differences in beef production in Ecuador into view. His research documents the ways that globalization of the meat trade and racial-ethnic land tenure differences shape land-use practices among smallholder cattle owners in the Ecuadorian Amazon. While both Amerindians and mestizos have seen a decline in the size of their landholdings, more Amerindians have found themselves in a position of needing to rent their lands in order to generate an income as cattle prices have gone through extreme fluctuations over the past two decades of global trade. For the lands that are rented, there are much higher rates of land degradation. Through Rudel's case study, we see how indigenous people's vulnerabilities to market fluctuations contribute to environmental degradation. With decreasing productivity of degraded lands, rearing cattle on these lands will be more difficult, thereby further exacerbating income inequality for Amerindians. Moving from production to processing, Freshour demonstrates how racial inequalities are used to the benefit of maintaining low wages by slaughterhouse managers.

Denny's chapter 7 and Chiles and Christy's chapter 8 highlight the complexity of thinking about the environment and smallholders in the context of a globalized food system. Denny's chapter "Contributions to Global Climate Change" focuses specifically on the greenhouse gas (GHG) emissions that come from animals' bodies and from their manure on the farm. She concludes that many places that are least efficient in GHG emissions are those most likely to be affected by climate change, yet least likely to be able to effectively respond to climate change impacts. While Denny's chapter focuses on a very specific type of GHG emission, she opens the door to thinking about what can be learned from diverse production settings that are more efficient and how policies and research might support lowering GHG emissions in animal agriculture.

Finally, Chiles and Christy's chapter "Livestock Intensification Strategies in Rwanda" offers an important intervention as it relates to thinking about humans' ethical obligations toward animals. Specifically, the authors focus on the development efforts that are currently underway to increase meat production in Rwanda, and more generally the global south, without any significant focus on animal suffering within intensive meat production systems and the ethical obligations humans have toward animals. Christy and Chiles seek to disrupt a development paradigm that views the growth of meat consumption as inevitable.

Confronting the Global Meat Industry—Possible Solutions

In the face of so many challenges confronting global meat production and consumption, it seems important to recognize that there is not total silence surrounding the costs of industrial meat production. There are spaces of resistance and opportunities for change. Table 9.1 provides examples of activities that individuals, organizations, and government policies have undertaken in an effort to curb or challenge the dominant system. In concluding, let us examine some of these proposals and evaluate how they may contribute to a better, more sustainable and just food system.

Consumers

Several initiatives for creating change in the global meat system fall within individual consumer activities, with occasional support from government initiatives. The decision to go vegetarian or vegan, opting to reduce meat

Table 9.1
Examples of attempts to create change in the global meat industry

Consumers	Organizational
Vegetarian/vegan diets	Animal welfare policies
Meatless Mondays	Cultured meat products/Clean meat
Purchasing local meat/eggs	Sustainability initiatives/Multi-stakeholder initiatives
Organic animal/products	Food sovereignty movement

consumption (e.g., Meatless Mondays), or choosing to purchase only meat or meat products sourced locally or grown organically are all examples of individual consumers trying to make a difference through their purchasing habits.

First, let us discuss vegetarian and vegan consumers. There are an estimated 1.5 billion vegetarians, or about 22 percent of the world's population (Leahy, Lyons, and Tol 2010, 4). Many of these people are not strict vegetarians, meaning that they occasionally eat meat; and some of them are vegetarian out of necessity because they lack the ability to purchase meat. Furthermore, vegetarianism is more common in some places than others. Most notably, more than 442 million people in India are vegetarian—that is more than one-third of the country's population, which is predominantly Hindu.[1] In other countries, there are far fewer vegetarians. In the United States, for example, only 5–6 percent of the population claims to be vegetarian (and 2 percent vegan), with those actually following a vegetarian diet estimated to be slightly less, anywhere from 1–3 percent of the population (Edelstein 2013; Maurer 2002; Newport 2012). Variation exists in Europe, where vegetarians make up about 3 percent of Portugal's population, while Italy and Germany have the highest number at approximately 9–10 percent of the population who identify as vegetarian (Micheletti and Stolle 2012). Most countries, then, have only a small share of the world's 1.5 billion vegetarians, and Asia—especially India—accounts for the majority of the world's vegetarians.

The reasons people give for becoming vegetarian vary. Most of the world's vegetarians adopted their diet because of religious beliefs (e.g., Hindu, Buddhist, and Seventh Day Adventist). Others identify health concerns or moral/ethical reasons, including environmental and animal welfare concerns, as their reasons for adopting a vegetarian diet. Vegans, those who opt out of using/consuming animal products completely, are overwhelmingly vegan for moral and ethical reasons (Maurer 2002; Micheletti and Stolle 2012). The motivation behind these choices is relevant in terms of thinking about how consumers' purchasing decisions, or what some call lifestyle politics, can create change in the global meat system. In the global north, there has been an increasing availability of vegetarian, plant-based options that are meant to serve as meat replacements. In this sense, consumers purchasing more vegetarian options have clearly encouraged companies to develop more of these product lines. In fact, even global meat corporations see this as a growth market: in 2016, Tyson became a minority investor in Beyond Meat, which produces plant-based alternatives to animal meat, and Tyson increased its ownership share in 2017 (Rowland 2017; Strom 2016). The role for lifestyle politics will be discussed further on, but let us first discuss a few other lifestyle choices, including individuals who opt to reduce their meat consumption.

Short of adopting a vegetarian and vegan diet, much of the global north has seen a trend of reduced meat consumption. For example, U.S. per capita meat consumption has decreased from 124.5 kg in 2004 to 113.9 kg in 2013 (FAO 2019), leading to a new category of consumers, named "flexitarians" (Flail 2011). Flexitarianism recognizes that people may not follow a strictly vegetarian diet, but they still consciously choose to reduce the amount of meat in their diets. Similarly, a movement that started in the United States in 2003 and has since gained a global presence is Meatless Mondays, a campaign to encourage people to go meatless for at least one day of the week (Monday Campaign 2017). Ghent, Belgium, made headlines in 2009 for declaring every Thursday in the city to be vegetarian, including all school cafeterias only serving vegetarian food for lunch (Bittman 2009; Harrell 2009). Despite the trend of increasing flexitarians and meatless days, it is important to note that most consumers opting to eat less meat are doing so in the context of living in countries with some of the highest rates of meat consumption in the world.

Local food initiatives have grown dramatically over the past decade in the United States, but also in other countries, including Japan (Kimura and Nishiyama 2007; Schupp 2016). "Food with a face," a phrase often associated with the local food movement, provides a double entendre when it comes to purchasing local meat. As a consumer participating in a local food movement, the ideal is not only that a consumer will know the farmer from which meat or meat products are purchased, but also that the farmer knows the animal, perhaps so much so that the animal has a name, from which a customer's steak or eggs originate.

Local meat consumption is seen as a mechanism to reduce some of the distance that global meat production imposes, thereby reducing some of the "bad" aspects of a long food chain. For example, for some consumers they gain a sense of trust by knowing the person who sells them their meat, thereby trying to guard against fears of meat harboring food-borne illnesses (see Gouveia and Juska 2002). For those concerned about animal welfare, there is also a feeling that locally produced animals lived better lives than industrially produced animals (see Weiss 2012). Furthermore, there are environmental ethics associated with locally sourced meats, including restaurant chefs who offer snout-to-tail delicacies as a means to promote less food waste and the idea that eating diverse breeds of animals not found in industrial settings encourages genetic diversity (Weiss 2012). Of course, what is conceived of as "local" is highly variable among consumers, and there is no guarantee to consumers that local production is inherently safer or more animal welfare–friendly, and like so many other consumer movements, there tend to be inequalities embedded in who participates and who does not or cannot participate (Allen 2004). Nonetheless, focusing on local sources is unique in that people not only "willingly inconvenience themselves but also they do so with deepening joy and increasingly significant effects" (Starr 2010, 487).

Organic meats and, more generally, all organic foodstuffs is a final area that has seen dramatic growth in terms of consumer purchasing in the last decade. Europe and North America lead the way in terms of purchasing of organic food and drink, which by 2014 had grown to $80 billion dollars annually, up from $15 billion in 1999 (Sahota 2016). The reasons consumers give for increased purchasing of organic products is similar to motivations for pursuing more local food production: that they lack trust in the safety and integrity of the food system; they believe there are health benefits to consuming organic products; and they are concerned about the environmental consequences of industrial agriculture. Meat remains a smaller part of the total organic market share in Europe than fruits and vegetables, with the exception of eggs and milk, which have made the greatest inroads in organic sales (Willer and Schaack 2016). In the United States, eggs, milk, and broiler chickens are among the top earners of organic meat products (Haumann 2017).

Small-p Politics

The four approaches already discussed—vegetarianism, meat-reduced diets, purchasing local meat products, and purchasing organic meats—fall under what many see as small-p politics (Kennedy, Johnston, and Parkins 2017), or what we call *smallitics* for short, which signifies consumers demonstrating their beliefs and values through their purchasing choices. Under debate with regard to consumer purchasing is whether or not these shopping decisions assist in creating an alternative food system, or simply serve to reinforce a food system that places the burden of change on individuals, with very little transformation of the dominant structures (e.g., corporate concentration; a subsidy system that favors corn and soybeans).

In the case of vegetarianism, scholars have noted that individuals who adopt a vegetarian lifestyle because of perceived health benefits often fail to engage an agenda of transformation of the food system, as these individuals may pay very little attention to broader political concerns related to the environment or animal welfare (Maurer 2002). In addition, despite a long history of vegetarians throughout the world, it remains unclear if vegetarian/vegan advocacy can diminish the growth of meat consumption globally (Neo and Emel 2017), in part because of the power and reach of the global meat industry.

In the case of Meatless Mondays, the movement has been accused of blunting vegetarianism's social change potential by adapting to a dominant culture that protects meat-eating culture against its long-perceived threats, including feminism,[2] animal rights and environmentalism (Singer 2017). Moreover, Meatless Mondays as a movement omits reference to nonhuman animals and the suffering they endure within industrial agricultural systems (Singer 2017). More broadly, Singer (2017) argues that small, incremental movements for change, like meat reduction movements, ignore important issues. These include the ways in which these campaigns frame why people should join (generally offering individualistic, nonradical logics); the fact that these movements do not move people toward total vegetarianism; and furthermore, how these movements rarely acknowledge that while some people choose to "do something" to improve society, their choice may do nothing for others in that society.

In the case of organics, the area that has seen the largest growth in consumer spending in terms of alternatives to the industrial agriculture, the reality is that organic production has increasingly conformed to or become a part of the industrial agricultural system (Guthman 2004; Neo and Emel 2017). While qualifying as "organic" production by meeting USDA organic regulations, the degree to which organic production varies from industrial production is questionable, given the sheer size of many of the organic farms and news reports that occasionally make headlines related to animal or worker cruelty on organic farms and slaughterhouses (e.g., Associated Press 2009). For these reasons, animal rights groups have pushed for stricter government regulation of animal welfare standards on organic farms (Associated Press 2017).

Despite all of these concerns, many scholars have argued it would be shortsighted to not recognize the potential for change in smallitics (Kennedy, Johnston, and Parkins 2017). Already, smallitics has made a difference in terms of increasing demand for plant-based meat replacements and organics, thereby incentivizing meat companies to invest in plant-based manufacturing companies and producers to switch to organic production. Similarly, small-scale producers all over the world, but especially in North America and Europe, have seen the revival of the appreciation for artisan production methods and products. Such smallitics have clearly made a difference in the lives of some humans and animals. However, the global meat industry seems persistent in its growth despite these changes. Consequently, we need to move beyond the level of consumer choice and smallitics to find more adequate solutions.

Consider again for a moment the case of India. With increases in meat production (from 1 MMT to 4 MMT) and per capita income (from US$4,200 to US$10,700) over the past twenty-five years, we might expect that meat consumption would have likewise increased. But meat consumption has not increased. Why not? At least part of the answer has to do with the reason why there are so many vegetarians in India, specifically religious beliefs and government policies that limit meat consumption. In other words, factors beyond the level of individual choice shape the diets of people in India in ways that counter the expanding meat industry. Thus, any search for responses to the global meat industry should consider the broader context of governmental policies and corporate or NGO activities.

Organizational Responses

Another effort to reform meat production has occurred in the form of animal welfare policies. While official government policies date back to the 1960s and 1970s (e.g., U.S. Humane Methods of Slaughter Act of 1978; the United Kingdom's Farm Animal Welfare Council of 1979), there has been a significant increase in animal welfare policies enacted by government and private organizations beginning in the 1990s and continuing to the present (Ransom 2007). For many, a turning point was the Treaty of Amsterdam that the EU signed in 1997, which recognized animals as sentient and required that EU member states consider animal welfare policies related to agriculture, transport, and research (Hirsch 2003).

Major corporations, including not only production and processing companies, such as Tyson Foods and Cargill, but also fast food companies, like McDonald's and KFC (as part of Yum! Brands) have adopted animal welfare standards in the past two decades. The shift in private organizations focusing on animal welfare policies can be considered in light of two factors. First, there has been an increase in government policies—not only national and regional (e.g., EU) level policies, but also international policies, such as the World Organization for Animal Health's (OIE) adoption of agricultural animal welfare guidelines in 2005. As a reference body to the World Trade Organization, the OIE is the first global governance organization that has provided guidance on animal welfare policies. Second, corporations are responding to the growing evidence of an increase in consumer demand for animal welfare policies across many Western, industrialized countries (Cornish, Raubenheimer, and McGreevy 2016). Of course, it is noteworthy that animal welfare standards do not address the labor conditions that can be found in CAFOs and processing facilities all around the globe.

In terms of actors contributing to change in the global meat system, a slightly different direction appears in the rise of cultured meat companies, that is, companies that culture meat in a lab using cellular science and in-vitro cultivation (Chiles 2013). The stuff of science fiction in the not too distant past, cultured meats increasingly are gaining mainstream appeal. Some consider cultured meat, also referred to as "clean meat" products, to be a potential solution to many of the concerns associated with global meat production. Not to be confused with plant-based meat substitutes, clean meats focus on growing meat in a lab environment using cellular tissue from animals. While scientific activity related to cultured meat has been underway since the 1970s, only in the 2000s has there been a major uptick in interest shown by major meat companies and prominent investors; the latter include Bill and Melinda Gates, who recently invested in cultured meat start-up companies (see Kowitt 2017). Then there are plant-based meat substitutes, like vegetarian burgers. Recently picked up by major news outlets was the recent U.S. Food and Drug Administration's approval of a vegetarian burger that bleeds (see Troitino 2018a).

Because cultured meat is one of the least-developed alternatives for creating change to our global meat system, it is difficult to assess the extent to which it would reduce or resolve some of the concerns raised in this volume. The reasons for the lack of certainty include technical concerns, in that cultured meat is in its infancy. It isn't clear yet whether these meats will be similar in taste, texture, safety, and nutrition to conventional meats, or if scaling up cultured meat production is economically feasible (Bonny et al. 2015; Kadim et al. 2015). Then there are the social, political, and environmental issues raised. Many see cultured meat as a mechanism to reduce animal suffering and resolve animal welfare concerns (Bonny et al. 2015), as well as environmental concerns, especially GHG emissions (Tuomisto and Mattos 2011). However, it remains to be seen if consumers are willing to accept cultured meat (Chiles 2013; Verbeke et al. 2015), and one area that is not addressed by alternative meats is corporate control of the food supply. Unless alternative ownership structures are explored, such as public-private partnerships, the amount of capital investment needed for the creation of cultured meat does not allow for a large number of laboratories involved in its production, thereby ensuring concentration of ownership.

The prospect of lab-created meat has many proponents and opponents, not least of which are animal farmers and farming-related interest groups in industrialized countries (Troitino 2018b). In part the criticism is a concern that the label "clean meat" disparages traditional sources of meat and can be misleading for consumers (AgFunder 2018; Troitino 2018b). Moreover, the rise of lab-grown meat furthers the argument made by the La Via Campesina landworkers alliance about the larger global agrifood system, which is that we are approaching an era of farmer-less agriculture that will only increase hunger and poverty (La Via Campesina 2009). As mentioned in the introduction to part III, we have seen a decline in the number of farmers in industrialized countries, as medium-sized farms are consolidated into larger farms.

La Via Campesina as a movement originates out of the global south and calls for peasants (smallholders) throughout the world to unite and fight for food sovereignty. Food sovereignty is defined as "the right of peoples to healthy and culturally appropriate food produced through ecologically sound and sustainable methods, and their right to define their own food and agriculture systems. It puts the aspirations and needs of those who produce, distribute and consume food at the heart of food systems and policies rather than the demands of markets and corporations" (Nyéléni International Steering Committee 2007).

To be clear: food sovereignty as a movement (first presented by La Vía Campesina in 1996 at a World Food Summit) and an organizing concept (the term "food sovereignty" was adopted by the Food and Agriculture Organization of the United Nations in 2012), makes no specific claims about meat production or consumption. Rather, in reviewing the definition of "food sovereignty" and the six pillars of food sovereignty, it is clear that the dominant, global meat system runs counter to achieving food sovereignty (Nyéléni International Steering Committee 2007). Furthermore, food sovereignty is one of the few efforts for creating change in the food system that incorporates smallholders across the globe. The previous examples of other efforts to create change in the global meat industry are largely enacted by people and organizations in Western, industrialized food landscapes, with very little real attention given to smallholder producers.

Finally, another means of reform is through sustainability initiatives that involve participants from all over the globe, although these initiatives generally represent corporate entities. Much of the effort to define and operationalize agricultural sustainability is now taking place in private settings, most notably multi-stakeholder initiatives (MSIs) (Loconto and Fouilleux 2014). MSI sustainability initiatives form to establish standards and certification that others must follow if they want their product to be deemed sustainable. However, these MSIs have to work to gain governance legitimacy, which means they must fend off possible criticism and competitors (Ponte 2014).

One of the earliest examples of an MSI for meat is the Marine Stewardship Council (MSC), established in 1999. A more recent example is the Aquaculture Stewardship Council (ASC), established in 2010. MSC focused on capture fisheries and sustainable fish, while ASC was created to focus on farm-raised fish. Ponte (2014, 264) in his astute analysis of sustainability

MSIs notes that *less* successful MSIs tend to be "more participatory, transparent and adopt more democratic and complex procedures," whereas MSC is an example of a successful MSI because it was more top-down in organizing, with more buy-in from major corporate entities involved in the industry. However, MSC was less successful at predicting and therefore incorporating governance of capture fisheries in the global south.

In the context of terrestrial animals, industry and nongovernmental organizations in the private sector are jointly organizing a new global sustainable beef initiative that seeks to reduce the environmental harms caused by beef production. The initiative, Global Roundtable on Sustainable Beef (GRSB), focuses on multiple regions of the world. The GRSB mission is "to advance continuous improvement in sustainability of the global beef value chain through leadership, science and multi-stakeholder engagement and collaboration" (GRSB n.d.).

The founding members of GRSB include Walmart, Solidaridad, the World Wildlife Fund, Cargill, Elanco, JBS, McDonald's, and Merck Animal Health. GRSB founders represent a reasonable degree of market embeddedness. Based on analysis of previous sustainability MSIs, the academic literature suggests several things about the future development and outcomes of GRSB (Ponte 2014; Schouten, Leroy, and Glasbergen 2012). First, GRSB is likely to have a market impact, given its members include some of the biggest purchasers of beef. Yet, the final indicators, metrics, or practices are not likely to represent a wide range of producers' interests, especially those of smallholders. A similar critique has also been leveled against ASC, in that smallholder aquaculture producers are not likely to benefit from its sustainability initiatives. In the case of GRSB, while the MSI's membership is more global than previous roundtables, its geographical representation remains fairly lopsided, with only a few members having any ties to Africa, and no membership affiliation for Asia. Finally, GRSB is likely to be similar to other roundtables in forgoing radical approaches, opting instead for pragmatic solutions to environmental problems (Schouten, Leroy, and Glasbergen 2012).

Climate Change and the Consequences of Business as Usual
While there are many alternative food movements underway to change or counter the global meat industry, it is also possible that the biggest change agent for the global meat industry moving forward will be extreme

weather-related events due to climate change. The scientific data is stacking up that we are on the cusp of an ecological tipping point, with hotter temperatures and more extreme weather events becoming the new normal (see Cribb 2010; Kolbert 2014; Rosenzweig et al. 2001; Samenow 2018). These events not only affect producers and consumers of meat products, they also impact the animals themselves. For countries in the global south, the FAO estimates an upward trend in the occurrence of natural disasters from 1980 to 2016, with flooding, drought, and other climatological disasters (extreme temperatures) accounting for the majority of agriculture losses (FAO 2017a). When focused on more industrialized production, the list is long of the environmental and human and animal welfare concerns associated with the operation of CAFOs. Less discussed are the ways in which CAFOs lack resiliency in the face of extreme weather events. The many issues include the problem of manure lagoons contaminating water supplies when flood waters breech their banks; a lack of adequate feed or water in the midst of drought; animals burned alive when the structures holding them catch fire; and animal deaths from drowning during flooding. In the United States, flooding from hurricanes have contributed to an inconceivable number of agricultural animal deaths. For example, 3.4 million chickens and 5,500 hogs were reported drowned during Hurricane Florence in 2018 in the North Carolina and in 2017 Hurricane Harvey was estimated to have killed thousands of cattle in the state of Texas, with total head counts not known, but loss estimates of around $93 million, including structures that housed the animals (Associated Press 2018; Fannin 2017). Finally, animal diseases, which have always plagued industrial animal production, are expected to become more of a threat as temperatures rise (Rosenzweig et al. 2001).

By now, it should be clear that there are a lot of different initiatives underway to create change—some small, some large—in our global meat system. What should also be evident is that each initiative has its limitations. There is no one-size-fits-all approach to redefining the global meat system. Yet, with overproduction and overconsumption situated alongside the impacts of climate change and the projected population growth reaching nine billion humans on the planet by 2040, very real concerns exist about our capacity to identify more efficient, humane, and ecologically sustainable food systems.

Notes

1. Depending in part on how a vegetarian diet is defined, estimates of the number of vegetarians vary widely, even in India where vegetarianism is relatively common. Leahy, Lyons, and Tol (2010, 4) offer a conservative estimate that 34 percent of Indians are vegetarian. Given a total population of 1.3 billion people, that would be about 442 million vegetarians. Edelstein (2013, 281), by contrast, states that about 80 percent of India's population "is believed to eat mostly a plant-based or vegetarian diet." This would place the number of vegetarians in India at about 1 billion, more than the rest of the world combined.

2. Feminists, particularly eco-feminists, have argued that consumption of meat is associated with the dominant form of masculinity and signifies a need to dominate not only animals and nature, but also other more marginalized humans (i.e., women, particularly racial-ethnic minority women).

References

Africanews. 2017. "South Africa's Poultry Industry Protests against EU Dumping." February 2. http://www.africanews.com/2017/02/02/south-africa-s-poultry-industry-protests-against-eu-dumping/. Accessed on February 15, 2019.

AgFunder. 2018. "It Shouldn't Be Called 'Clean Meat' But It Will Be Clean, Says Cultured Meat's Founder." *Successful Farming*, July 27. https://www.agriculture.com/news/technology/it-shouldn-t-be-called-clean-meat-but-it-will-be-clean-says-cultured-meat-s-founder. Accessed on July 28, 2018.

Albert, James A. 1991. "A History of Attempts by the Department of Agriculture to Reduce Federal Inspection of Poultry Processing Plants—A Return to the Jungle." *Louisiana Law Review* 51 (6): 1183–1231.

Alexandratos, N., and J. Bruinsma. 2012. "World Agriculture towards 2030/2050: The 2012 Revision." ESA Working Paper No. 12-03, June. Rome: Food and Agriculture Organization of the United Nations.

Allen, Patricia. 2004. *Together at the Table: Sustainability and Sustenance in the American Agrifood System*. University Park: Pennsylvania State University Press.

Alliance for the Great Lakes. 2017. "Asian Carp Threat." August 31. https://greatlakes.org/2017/08/asian-carp-threat/. Accessed on February 12, 2019.

An, Tong-Qing, Zhi-Jun Tian, Chao-Liang Leng, Jin-Mei Peng, and Guang-Zhi Tong. 2011. "Highly Pathogenic Porcine Reproductive and Respiratory Syndrome Virus, Asia." *Emerging Infectious Diseases* 17 (9): 1782–1784.

Anderson, Anthony, Peter May, and Michael Balick. 1991. *The Subsidy from Nature: Palm Forests, Peasantry, and Development on an Amazon Frontier*. New York: Columbia University Press.

Anderson, E. N. 1988. *The Food of China*. New Haven, CT: Yale University Press.

Andersson, M. S., A. Saltzman, P. S. Virk, and W. H. Pfeiffer. 2017. "Progress Update: Crop Development of Biofortified Staple Food Crops under HarvestPlus." *African Journal of Food, Agriculture, Nutrition and Development* 17 (2): 11905–11935.

Asche, Frank, Marc F. Bellemare, Cathy Roheim, Martin D. Smith, and Sigbjørn Tveteras. 2015. "Fair Enough? Food Security and the International Trade of Seafood." *World Development* 67:151–160.

Associated Press. 2008. "Poultry Plants Are Raided." *Los Angeles Times*, April 17. http://articles.latimes.com/2008/apr/17/nation/na-raid17. Accessed on February 15, 2019.

Associated Press. 2009. "Vermont Slaughterhouse Closed amid Animal Cruelty Allegations." *Los Angeles Times*, November 3. http://latimesblogs.latimes.com/unleashed/2009/11/vermont-slaughterhouse-closed-amid-animal-cruelty-allegations.html. Accessed on September 21, 2017.

Associated Press. 2017. "Ag Dep't Delays Animal Welfare Standards for Organic Meats." *US News*, May 9. https://www.usnews.com/news/politics/articles/2017-05-09/ag-dept-delays-animal-welfare-standards-for-organic-meats. Accessed on July 15, 2018.

Associated Press. 2018. "3.4 Million Chickens, 5,500 Hogs Dead in Florence Flood." *PBS NewsHour*, September 19. https://www.pbs.org/newshour/nation/3-4-million-chickens-5500-hogs-dead-in-florence-flood. Accessed on September 20, 2018.

Aued, B. 2007. "Unions Work to Lure Hispanics." *Athens Banner Herald*, September 2. http://onlineathens.com/stories/090207/news_20070902062.shtml#.Wb7bdrJ97mE. Accessed on April 13, 2016.

Bacon, David. 2012. "How US Policies Fueled Mexico's Great Migration." *The Nation*, January 4, 2012. https://www.thenation.com/article/how-us-policies-fueled-mexicos-great-migration/. Accessed on February 4, 2012.

Bailey, Conner. 1988. "Social Consequences of Tropical Shrimp Mariculture Development." *Ocean and Shoreline Management* 11:31–44.

Barabási, Albert-László, and Eric Bonabeau. 2003. "Scale-Free Networks." *Scientific American* 288 (5): 60–69.

Baran, Paul A., and Paul M. Sweezy. 1966. *Monopoly Capital: An Essay on the American Economic and Social Order*. New York: Monthly Review Press.

Barboza, David, and Andrew Ross Sorkin. 2001. "Tyson to Acquire IBP in $3.2 Billion Deal." *New York Times*, January 2. http://www.nytimes.com/2001/01/02/us/tyson-to-acquire-ibp-in-3.2-billion-deal.html. Accessed on February 14, 2017.

Barnes, Cindy Brown, and Steve Morris. 2016. *Workplace and Safety Health: Additional Data Needed to Address Continued Hazards in the Meat and Poultry Industry*. GAO: U.S. Government Accountability Office. https://www.gao.gov/assets/680/676796.pdf.

Bartels, Larry M. 2010. *Unequal Democracy: The Political Economy of the New Gilded Age*. Princeton, NJ: Princeton University Press.

References

Bavier, Joe. 2018. "SA Farmers Play 'Chicken' with US Tariffs." *AGOA.Info*, September 29. https://agoa.info/news/article/15513-sa-farmers-play-chicken-with-us-tariffs.html. Accessed on October 1, 2018.

Beardsworth, Alan, and Teresa Keil. 1992. "The Vegetarian Option: Varieties, Conversions, Motives and Careers." *The Sociological Review* 40 (2): 253–293.

Beck, Nathaniel, and Jonathan N. Katz. 1995. "What to Do (and Not to Do) with Time-Series Cross-Section Data." *The American Political Science Review* 89 (3): 634–647.

Beck, Nathaniel, and Jonathan N. Katz. 1996. "Nuisance vs. Substance: Specifying and Estimating Time-Series-Cross-Section Models." *Political Analysis* 6: 1–36.

Beck, Nathaniel, and Jonathan N. Katz. 2004. "Time-Series—Cross-Section Issues: Dynamics 2004." Working paper. http://www.nyu.edu/gsas/dept/politics/faculty/beck/beckkatz.pdf.

Bell, David E., and Catherine Ross. 2008. *Case Study: JBS Swift & Co.* N9-509-021. Cambridge, MA: Harvard Business School.

Belton, Ben, Simon R. Bush, and David C. Little. 2018. "Not Just for the Wealthy: Rethinking Farmed Fish Consumption in the Global South." *Global Food Security* 16:85–92.

Bett, Hillary Kiplangat. 2012. "Economic Analysis of Indigenous Chicken Genetic Resources in Kenya." PhD. diss., Humboldt University, Berlin.

Bittman, Mark. 2008. "Rethinking the Meat-Guzzler." *New York Times*, January 27. http://www.nytimes.com/2008/01/27/weekinreview/27bittman.html. Accessed on September 21, 2017.

Bittman, Mark. 2009. "Ghent Goes Vegetarian." *New York Times*, May 15. https://dinersjournal.blogs.nytimes.com/2009/05/15/ghent-goes-vegetarian/?mcubz=1. Accessed on September 21, 2017.

Bonanno, Alessandro, and Douglas H. Constance. 2010. *Stories of Globalization: Transnational Corporations, Resistance, and the State.* University Park: Penn State Press.

Bonny, S. P., G. E. Gardner, D. W. Pethick, and J. F. Hocquette. 2015. "What Is Artificial Meat and What Does It Mean for the Future of the Meat Industry?" *Journal of Integrative Agriculture* 14 (2): 255–263.

Booth, David, and Frederick Golooba-Mutebi. 2014. "Policy for Agriculture and Horticulture in Rwanda: A Different Political Economy?" *Development Policy Review* 32 (s2): s173–s196.

Bourette, Susan. 2009. *Meat: A Love Story: Pasture to Plate, a Search for the Perfect Meal.* New York: Penguin Press.

Bourne, Joel K., Jr. 2014. "How to Farm a Better Fish." *National Geographic*, June. http://www.nationalgeographic.com/foodfeatures/aquaculture/. Accessed on July 16, 2018.

Boyd, Josh. 2000. "Actional Legitimation: No Crisis Necessary." *Journal of Public Relations Research* 12 (4): 341–353.

Boyd, William, and Michael Watts. 1997. "Agro-Industrial Just-in-Time: The Chicken Industry and Post-War American Capitalism." In *Globalising Food: Agrarian Questions and Global Restructuring*, ed. Michael Goodman and David Watts, 192–225. London: Routledge.

Briggs, Matthew, Simon Funge-Smith, Rohana Subasighe, and Michael Phillips. 2004. *Introductions and Movement of Penaeus vannamei and Penaeus stylirostris in Asia and the Pacific*. RAP Publication 2004/10. Bangkok: Regional Office for Asia and the Pacific, Food and Agriculture Organization of the United Nations. http://www.fao.org/docrep/007/ad505e/ad505e00.htm. Accessed on January 20, 2019.

Brown, Michael J. 2017. *Petition to Permit Waivers of the Maximum Line Speed Rates for Young Chicken Slaughter Establishments under the New Poultry Inspection System and Salmonella Initiative Program*. Washington, DC: National Chicken Council. https://www.fsis.usda.gov/wps/wcm/connect/7734f5cf-05d9-4f89-a7eb-6d85037ad2a7/17-05-Petition-National-Chicken-Council-09012017.pdf?MOD=AJPERES. Accessed on January 20, 2019.

Brown, Warren. 1979. "Union Fighting Southern Traditions as Well as Mississippi Poultry Plant." *Washington Post*, December 2. https://www.washingtonpost.com/archive/politics/1979/12/02/union-fighting-southern-traditions-as-well-as-mississippi-poultry-plant/55e07dab-c7b1-43fd-9e1e-6686da542495/?noredirect=on&utm_term=.00557cdf9742. Accessed on June 14, 2014.

Buller, Henry, and Carol Morris. 2003. "Farm Animal Welfare: A New Repertoire of Nature-Society Relations or Modernism Re-embedded?" *Sociologia Ruralis* 43 (3): 216–237.

Bureau of the Census, U.S. Department of Commerce. 1990. *General Population Characteristics: Georgia*. Washington, DC: Government Printing Office. https://www2.census.gov/library/publications/decennial/1990/cp-1/cp-1-12.pdf#. Accessed on January 20, 2019.

Busch, Lawrence. 2008. "Nanotechnologies, Food, and Agriculture: Next Big Thing or Flash in the Pan?" *Agriculture and Human Values* 25 (2): 215–218.

CAADP (Comprehensive Africa Agriculture Development Programme). 2013. *Nutrition Country Paper—Rwanda*. East and Central Africa Regional CAADP Nutrition Program Development Workshop. Rome: UN-FAO. http://www.fao.org/fileadmin/user_upload/nutrition/docs/policies_programmes/CAADP/east_central_africa/outputs/country_papers/Rwanda_NCP_210213.pdf. Accessed on January 28, 2019.

References

Calle, Alicia, Florencia Montagnini, and Andres Zuluaga. 2009. "Farmers' Perceptions of Silvopastoral System Promotion in Quindio, Colombia." *Bois et Forets des Tropiques* 300:79–94.

Calva, Alejandra. 2016. *The Hispanic / Latinx Community of Athens-Clarke County, Ga. in 2016: A Comprehensive Needs Assessment and Recommendations for Service Providers*. Athens: Latin American & Caribbean Studies Institute, University of Georgia. https://lcfgeorgia.org/wp-content/uploads/2017/01/acc-latinx-needs-assessment-report-final-11.pdf.

Campbell, Brooke, and Daniel Pauly. 2013. "Mariculture: A Global Analysis of Production Trends Since 1950." *Marine Policy* 39:94–100.

Campbell, T. Colin, and Thomas M. Campbell II. 2006. *The China Study*. Dallas: Benbella Books.

Carolan, Michael. 2005. "Barriers to the Adoption of Sustainable Agriculture on Rented Land: An Examination of Contesting Social Fields." *Rural Sociology* 70:387–413.

Carolan, Michael. 2012. *The Sociology of Food and Agriculture*. New York: Routledge.

Carolan, Michael. 2013. *The Real Cost of Cheap Food*. 2nd ed. New York: Routledge.

Carolan, Michael. 2014. *Cheaponomics: The High Cost of Low Prices*. New York: Routledge.

Carriere, Stephanie M., Mathieu Andre, Philippe Letourmy, Isabelle Olivier, and Doyle B. McKey. 2002. "Seed Rain Beneath Remnant Trees in a Slash-and-Burn Agricultural System in Southern Cameroon." *Journal of Tropical Ecology* 18:353–374.

Carvajal, Doreen, and Stephen Castle. 2009. "A U.S. Hog Giant Transforms Eastern Europe." *New York Times*, May 6, sec. Business/Global business. http://www.nytimes.com/2009/05/06/business/global/06smithfield.html. Accessed on September 7, 2014.

Censo del Ecuador. 2011. "Resultados del Censo 2010 de Población y Vivienda en el Ecuador." http://www.ecuadorencifras.gob.ec//wp-content/descargas/Manu-lateral/Resultados-provinciales/morona_santiago.pdf. Accessed on November 18, 2017.

Chang, K. C., ed. 1977. *Food in Chinese Culture*. New Haven, CT: Yale University Press.

Chase-Dunn, Christopher, Yukio Kawano, and Benjamin D. Brewer. 2000. "Trade Globalization Since 1795: Waves of Integration in the World-System." *American Sociological Review* 65 (1): 77–95.

Chen, Jia-Ching, John Aloysius Zinda, and Emily Ting Yeh. 2017. "Recasting the Rural: State, Society and Environment in Contemporary China." *Geoforum* 78 (1): 83–88.

Chiles, R. M. 2013. "Intertwined Ambiguities: Meat, in vitro Meat, and the Ideological Construction of the Marketplace." *Journal of Consumer Behavior* 12:472–482.

China Pollution Source Census. 2010. *Zhongguo Wuran Yuan Pucha* [*China Pollution Source Census*]. http://cpsc.mep.gov.cn/gwgg/. Accessed on December 2, 2018.

CIA (Central Intelligence Agency). 2016. *World Factbook Book: Rwanda*. https://www.cia.gov/library/publications/the-world-factbook/geos/rw.html. Accessed online December 2017.

Clapp, Jennifer. 2003. "Transnational Corporate Interests and Global Environmental Governance: Negotiating Rules for Agricultural Biotechnology and Chemicals." *Environmental Politics* 12 (4): 1–23.

Clapp, Jennifer, and Doris Fuchs, eds. 2009. *Corporate Power in Global Agrifood Governance*. Cambridge, MA: MIT Press.

Clark, Colin W. 1973. "The Economics of Overexploitation." *Science* 181:630–634.

Clarke, William. 2015. "JBS Pulled into Brazil's Anti-Corruption Crackdown." Agrimoney.com, November 26. http://www.ellinghuysen.com/article/20151130_jbs_corruption_investigation/ellinghuysen. Accessed on February 14, 2017.

Cochrane, D. T. 2010. "Review of Nitzan and Bichler's 'Capital as Power: A Study of Order and Creorder.'" *Theory in Action* 3 (2): 110–116.

Cody, Edward. 2007. "Oh, to Be Born in the Year of the Pig." *Washington Post*, March 1. http://www.washingtonpost.com/wp-dyn/content/article/2007/02/28/AR2007022802104.html. Accessed on May 13, 2018.

Compa, Lance A. 2004. *Blood, Sweat, and Fear: Workers' Rights in U.S. Meat and Poultry Plants*. Human Rights Watch. https://www.hrw.org/reports/2005/usa0105/. Accessed on March 19, 2018.

Conant, Richard T., and Keith Paustian. 2002. "Potential Soil Carbon Sequestration in Overgrazed Grassland Ecosystems." *Global Biogeochemical Cycles* 16 (4): 1143. https://doi.org/10.1029/2001GB001661.

Constance, Douglas H., and William D. Heffernan. 1991. "The Global Poultry Agro/Food Complex." *International Journal of Sociology of Agriculture and Food* 1:126–142.

Constance, Douglas H., Francisco Martinez, and Gilberto Aboites. 2010. "The Globalization of the Poultry Industry: Tyson Foods and Pilgrim's Pride in Mexico." In *From Community to Consumption: New and Classical Themes in Rural Sociological Research*, vol. 16, ed. Alessandro Bonanno, 59–75. Bingley: Emerald Group Publishing.

Constance, Douglas H., Francisco Martinez-Gomez, Gilberto Aboites-Manrique, and Alessandro Bonanno. 2013. "The Problems with Poultry Production and Processing." In *The Ethics and Economics of Agrifood Competition*, ed. Harvey S. James, 155–176. New York: Springer.

Cornish, A., D. Raubenheimer, and P. McGreevy. 2016. "What We Know about the Public's Level of Concern for Farm Animal Welfare in Food Production in Developed Countries." *Animals* 6 (11): 74. https://doi.org/10.3390/ani6110074. Accessed on February 2, 2019.

Cribb, Julian. 2010. *The Coming Famine: The Global Food Crisis and What We Can Do to Avoid It*. Berkeley: University of California Press.

Dagang, Alyson, and P. Nair. 2003. "Silvopastoral Research and Adoption in Central America: Recent Findings and Recommendations for Future Directions." *Agroforestry Systems* 59 (2): 149–155.

Davison, Nicola. 2013. "Rivers of Blood: The Dead Pigs Rotting in China's Water Supply." *Guardian*, March 29, sec. World news. https://www.theguardian.com/world/2013/mar/29/dead-pigs-china-water-supply. Accessed on February 20, 2019.

Day, Alexander. 2013. *The Peasant in Postsocialist China: History, Politics, and Capitalism*. Cambridge, UK: Cambridge University Press.

Day, Sherri. 2003. "Jury Clears Tyson Foods in Use of Illegal Immigrants." *New York Times*, March 27. https://www.nytimes.com/2003/03/27/us/jury-clears-tyson-foods-in-use-of-illegal-immigrants.html. Accessed on February 14, 2019.

De Barcellos, Marcia Dutra, Klaus Grunert, Yangeng Zhou, Wim Verbeke, F. J. A. Perez-Cueto, and Athanasios Krystallis. 2012. "Consumer Attitudes to Different Pig Production Systems: A Study from Mainland China." *Agriculture and Human Values* 30: 443–455.

De Soto, Hernan. 1989. *The Other Path: The Invisible Revolution in the Third World*. New York: Harper Collins.

Degan, Ronald Jean, and K. Matthew Wong. 2012. "An Examination of the Resource-Based Horizontal Acquisition Strategy of JBS—the Biggest Meat Packer in the World." *Proceedings of the New York State Economics Association*, vol. 5, ed. Richard Vogel, 37–46. Farmingdale, NY: New York State Economics Association.

Demment, Montague W., Michelle M. Young, and Ryan L. Sensenig. 2003. "Providing Micronutrients through Food-Based Solutions: A Key to Human and National Development." *The Journal of Nutrition* 133 (11): 3879S–3885S.

Devereux, Stephen. 2014. "Livestock and Livelihoods in Africa: Maximising Animal Welfare and Human Wellbeing." Working Paper No. 451. Brighton, UK: Institute of Development Studies.

Diaz, Robert J., and Rutger Rosenberg. 2008. "Spreading Dead Zones and Consequences for Marine Ecosystems." *Science* 321 (5891): 926–929.

Dierolf, Thomas S., Rienzzie Kern, Tim Ogborn, Mark Protti, and Marvin Schwartz. 2002. "Heifer International: Growing a Learning Organisation." *Development in Practice* 12 (3–4): 436–448.

Dietz, Thomas, and Eugene A. Rosa. 1997. "Effects of Population and Affluence on CO_2 Emissions." *Proceedings of the National Academy of Sciences of the United States of America* 94 (1): 175–179.

Dimsdale, Parks B. 1970. "A History of the Cotton Producers Association." PhD diss., University of Florida.

Domina, David A., and C. Robert Taylor. 2009. "The Debilitating Effects of Concentration Markets Affecting Agriculture." *Drake Journal of Agricultural Law* 15:61–108.

Done, Hansa Y., Arjun K. Venkatesan, and Rolf U. Halden. 2015. "Does the Recent Growth of Aquaculture Create Antibiotic Resistance Threats Different from Those Associated with Land Animal Production in Agriculture?" *American Association of Pharmaceutical Scientists Journal* 17 (3): 513–524.

Duggan, Jennifer. 2014. "Dead Pigs Floating in Chinese River." *Guardian*, April 17. https://www.theguardian.com/environment/chinas-choice/2014/apr/17/china-water. Accessed on March 20, 2017.

Easley, David, and Jon Kleinberg. 2010. *Networks, Crowds, and Markets: Reasoning about a Highly Connected World*. New York: Cambridge University Press.

Eaton, Emily. 2013. *Growing Resistance: Canadian Farmers and the Politics of Genetically Modified Wheat*. Winnipeg: University of Manitoba Press.

Economist. 2007. "The Golden Pig Cohort." February 8. http://www.economist.com/node/8681045. Accessed on March 19, 2014.

Economist. 2014. "Swine in China: Empire of the Pig." December 17. https://www.economist.com/christmas-specials/2014/12/17/empire-of-the-pig. Accessed on December 2, 2018.

Edelstein, Sari. 2013. *Food Science: An Ecological Approach*. 2nd ed. Burlington, MA: Jones & Bartlett Learning.

Emel, Jody, and Harvey Neo, eds. 2015. *Political Ecologies of Meat*. London: Routledge.

English, Philip, Patrick McSharry, and Kasim Ggombe. 2016. *Raising Exports and Attracting FDI in Rwanda*. International Growth Centre, November. Policy Brief 38402.

Ennis, Sharon R., Merarys Rios-Vargas, and Nora G. Alber. 2011. *The Hispanic Population: 2010*. C2010BR-04. Washington, DC: Government Printing Office. https://www.census.gov/prod/cen2010/briefs/c2010br-04.pdf. Accessed on December 15, 2018.

Esposito, Anthony. 2016. "Chile Court Orders Salmon Farms Antibiotic Use to Be Disclosed." *Reuters*, June 1. http://www.reuters.com/article/us-chile-environment-salmon-idUSKCN0YN5J9. Accessed on August 11, 2018.

Fannin, Blair. 2017. "Texas Agricultural Losses from Hurricane Harvey Estimated at More Than $200 Million." *AgriLife Today*, October 27. Texas A&M AgriLife. https://

References

today.agrilife.org/2017/10/27/texas-agricultural-losses-hurricane-harvey-estimated-200-million/. Accessed on September 20, 2018.

Fantasia, Rick, and Kim Voss. 2004. *Hard Work: Remaking the American Labor Movement*. Berkeley: University of California Press.

FAO (Food and Agriculture Organization of the United Nations). 2009. *The State of World Fisheries and Aquaculture*. https://www.scribd.com/document/18420227/FAO-The-State-of-World-Fisheries-and-Aquaculture. Accessed on August 11, 2018.

FAO (Food and Agriculture Organization of the United Nations). 2016. *The State of World Fisheries and Aquaculture*. http://www.fao.org/3/a-i5555e.pdf. Accessed on August 11, 2018.

FAO (Food and Agriculture Organization of the United Nations). 2017a. *The Impact of Disasters on Agriculture: Addressing the Information Gap*. Rome: FAO.

FAO (Food and Agriculture Organization of the United Nations). 2017b. *Food Outlook*. http://www.fao.org/giews/reports/food-outlook/en/. Accessed on December 15, 2018.

FAO (Food and Agriculture Organization of the United Nations). 2018a. *The State of Food Security and Nutrition in the World: Building Climate Resilience for Food Security and Nutrition*. Rome: FAO.

FAO (Food and Agriculture Organization of the United Nations). 2018b. *The State of World Fisheries and Aquaculture 2018*. http://www.fao.org/documents/card/en/c/I9540EN/. Accessed on August 11, 2018.

FAO (Food and Agriculture Organization of the United Nations). 2019. *FAOSTAT*. Food and agriculture data. http://www.fao.org/faostat/en/#home. Accessed on January 9, 2019.

FAO (Food and Agriculture Organization of the United Nations). n.d. *Fishery Statistics: Reliability and Policy Implications*. Undated statement by the FAO Fisheries Department. http://www.fao.org/docrep/FIELD/006/Y3354M/Y3354M00.HTM. Accessed on August 11, 2018.

Fearnside, Philip M. 2001. "Soybean Cultivation as a Threat to the Environment in Brazil." *Environmental Conservation* 28 (1): 23–38.

Fink, Leon. 2003. *The Maya of Morganton: Work and Community in the Nuevo New South*. Chapel Hill: University of North Carolina Press.

Firebaugh, Glenn. 2008. *Seven Rules for Social Research*. Princeton, NJ: Princeton University Press.

Flail, G. J. 2011. "Why 'Flexitarian' Was a Word of the Year: Carno-phallogocentrism and the Lexicon of Vegetable-Based Diets." *International Journal of Humanities and Social Science* 1 (12) (September): 83–92.

Fortson, Leni, and Joanna Hawkins. 2015. "Deficient Medical Management Leads to Musculoskeletal Injuries at Delaware Poultry Processing Plant." *OSHA Regional News Brief—Region 3*, September 2. Washington, DC: U.S. Department of Labor, Occupational Safety and Health Administration.

Foster, Brendan. 2018. "Australia Shocked by Death of 2,400 Sheep on Ship to Qatar." *New York Times*, April 9, 2018. https://www.nytimes.com/2018/04/09/world/australia/qatar-sheep-deaths.html. Accessed on February 5, 2019.

Frantz, Douglas. 1994. "How Tyson Became the Chicken King." *New York Times*, August 28. https://www.nytimes.com/1994/08/28/business/how-tyson-became-the-chicken-king.html. Accessed on April 28, 2018.

Fraser, David. 2008. "Toward a Global Perspective on Farm Animal Welfare." *Applied Animal Behaviour Science* 113 (4): 330–339.

Freitas, Gerson, Jr., Tatiana Freitas, and Jeff Wilson. 2017. "Dirty Family Secret Is behind JBS's $20 Billion Buying Spree." Bloomberg.com, May 26. https://www.bloomberg.com/news/articles/2017-05-26/the-dirty-family-secret-behind-jbs-s-20-billion-buying-spree. Accessed on May 26, 2017.

Freudenburg, William R. 2005. "Privileged Access, Privileged Accounts: Toward a Socially Structured Theory of Resources and Discourses." *Social Forces* 84 (1): 89–114.

Fritzsche, Tom. 2013. *Unsafe at These Speeds: Alabama's Poultry Industry and Its Disposable Workers*. Southern Poverty Law Center. https://www.splcenter.org/20130228/unsafe-these-speeds. Accessed on February 15, 2014.

Fry, Richard. 2008. *Latino Settlement in the New Century*. October. Washington, DC: Pew Hispanic Center.

Fu, C., W. Hu, Y. Wang, and Z. Zhu. 2005. "Developments in Transgenic Fish in the People's Republic of China." *Revue scientifique et technique (International Office of Epizootics)* 24 (1): 299–307.

Gale, Fred. 2017. *China's Pork Imports Rise Along with Production Costs*. Washington, DC: USDA Economic Research Service. https://www.ers.usda.gov/webdocs/publications/81948/ldpm-271-01.pdf?v=42745.

Gallet, Craig A. 2010. "The Income Elasticity of Meat: A Meta-Analysis." *Australian Journal of Agricultural and Resource Economics* 54 (4): 477–490.

Gao, Mark. 2012. "Meat Industry Is Big Business in China." *Global Meat*. http://www.globalmeatnews.com/Industry-Markets/Meat-industry-is-big-business-in-China. Accessed on May 14, 2018.

Garay, Anabelle. 2009. "Pilgrim's Pride Pays $4.5M to End Immigrant Probe." *Seattle Times*, December 30. http://old.seattletimes.com/html/businesstechnology/2010642321_apuspilgrimsprideimmigration.html. Accessed on February 15, 2019.

Georgia Office of Disability Adjudication and Review. 2018. https://www.disabilityjudges.com/state/georgia. Accessed on February 2, 2019.

Georgia Power. 2016. *Food Processing in Georgia: Georgia's Top Manufacturing Sector*. Georgia Power, Community & Economic Development, Atlanta. https://www.selectgeorgia.com/resources/publications/food-processing-georgia/. Accessed on February 2, 2019.

Gerber, Pierre J., Anne Mottet, Carolyn I. Opio, Alessandra Falcucci, and Félix Teillard. 2015. "Environmental Impacts of Beef Production: Review of Challenges and Perspectives for Durability." *Meat Science* 109 (May 20): 2–12.

Gewertz, Deborah, and Frederick Errington. 2010. *Cheap Meat: Flap Food Nations in the Pacific Islands*. Berkeley: University of California Press.

Ggombe, Kasim, and Richard S. Newfarmer. 2017. "Rwanda: From Devastation to Services-First Transformation." WIDER Working Paper 84.

Gilens, Martin. 2014. *Affluence and Influence: Economic Inequality and Political Power in America*. Princeton, NJ: Princeton University Press.

Gilmore, Ruth W. 2006. *Golden Gulag: Prisons, Surplus, Crisis, and Opposition in Globalizing California*. Berkeley: University of California Press.

Gisolfi, Monica R. 2017. *The Takeover: Chicken Farming and the Roots of American Agribusiness*. Athens: University of Georgia Press.

GiveDirectly. 2018. "Research at GiveDirectly." https://givedirectly.org/research-at-give-directly. Accessed on September 12, 2018.

GiveWell, 2018. "Gifts of Livestock Programs." https://www.givewell.org/international/economic-empowerment/livestock-gift-programs. Accessed on September 12, 2018.

Gjøen, H. M. and H. B. Bentsen. 1997. "Past, Present and Future of Genetic Improvement in Salmon Aquaculture." *ICES Journal of Marine Science* 54 (6): 1009–1014.

Glenn, Evelyn Nakano. 2002. *Unequal Freedom: How Race and Gender Shaped American Citizenship and Labor*. Cambridge, MA: Harvard University Press.

Glover, Richard. 2018. "The Shame of Australia's Live Sheep Trade." *Washington Post*, May 17. https://www.washingtonpost.com/news/global-opinions/wp/2018/05/17/the-shame-of-australias-live-sheep-trade/?utm_term=.588709f9db1e. Accessed on August 10, 2018.

Golash-Boza, Tanya, and Pierrette Hondagneu-Sotelo. 2013. "Latino Immigrant Men and the Deportation Crisis: A Gendered Racial Removal Program." *Latino Studies* 11 (3): 271–292. https://doi.org/10.1057/lst.2013.14. Accessed on February 2, 2019.

Gouveia, L., and A. Juska. 2002. "Taming Nature, Taming Workers: Constructing the Separation between Meat Consumption and Meat Production in the U.S." *Sociologia Ruralis* 42 (4): 370–390.

Grau, Ricardo, Tobias Kuemmerle, and Leandro Macchi. 2013. "Beyond 'Land Sparing versus Land Sharing': Environmental Heterogeneity, Globalization and the Balance between Agricultural Production and Nature Conservation." *Current Opinion in Environmental Sustainability* 5 (5): 477–483.

Gray, LaGuana. 2014. *We Just Keep Running the Line: Black Southern Women and the Poultry Processing Industry*. Baton Rouge: LSU Press.

Greene, Joel L. 2015. *Update on the Highly-Pathogenic Avian Influenza Outbreak of 2014–2015*. Washington, DC: Congressional Research Service.

Griffith, David. 1990. "Consequences of Immigration Reform for Low-Wage Workers in the Southeastern U.S.: The Case of the Poultry Industry." *Urban Anthropology and Studies of Cultural Systems and World Economic Development* 19 (1/2): 155–184.

Griffiths, D., P. Van Khanh, and T. Q. Trong. 2010. *Cultured Aquatic Species Information Programme, Pangasius hypophthalmus*. FAO Fisheries and Aquaculture Department, Food and Agriculture Organization of the United Nations. http://www.fao.org/fishery/culturedspecies/Pangasius_hypophthalmus/en. Accessed on August 11, 2018.

GRSB. n.d. "About GRSB." https://grsbeef.org/. Accessed on August 8, 2017.

Guèye, El Fallou. 2000. "The Role of Family Poultry in Poverty Alleviation, Food Security and the Promotion of Gender Equality in Rural Africa." *Outlook on Agriculture* 29 (2): 129–136.

Guo, Hongdong, Robert W. Jolly, and Jianhua Zhu. 2007. "Contract Farming in China: Perspectives of Farm Households and Agribusiness Firms." *Comparative Economic Studies* 49 (2): 285–312.

Guptill, Amy, and Rick Welsh. 2014. "The Declining Middle of American Agriculture: A Spatial Phenomenon." In *Rural America in a Globalizing World*, ed. Conner Bailey, Leif Jenson, and Elizabeth Ransom, 36–50. Morgantown: West Virginia University Press.

Guthey, Greig. 2001. "Mexican Places in Southern Spaces: Globalization, Work, and Daily Life in and around the North Georgia Poultry Industry." In *Latino Workers in the Contemporary South*, ed. A. D. Murphy, C. Blanchard, and J. A. Hill, 57–67. Athens: University of Georgia Press.

Guthman, J. 2004. *Agrarian Dreams: The Paradox of Organic Farming in California*. Berkeley: University of California Press.

References

Hallegatte, Stephane, Mook Bangalore, Laura Bonzanigo, Marianne Fay, Tamaro Kane, Ulf Narloch, Julie Rozenberg, David Treguer, and Adrien Vogt-Schilb. 2016. *Shock Waves: Managing the Impacts of Climate Change on Poverty*. Washington DC: The World Bank. https://openknowledge.worldbank.org/bitstream/handle/10986/22787/9781464806735.pdf. Accessed on February 15, 2019.

Halverson, Nathan. 2015. "How China Purchased a Prime Cut of the Pork Industry." *Reveal*, January 24. https://www.revealnews.org/article/how-china-purchased-a-prime-cut-of-americas-pork-industry/. Accessed on May 14, 2018.

Hamerschlag, Kari. 2018. "Leading NGOs Expose Greenwashing by U.S. Roundtable for Sustainable Beef, Demand Real Sustainability Plan." *Friends of the Earth*. https://foe.org/news/leading-ngos-expose-greenwashing-u-s-roundtable-sustainable-beef-demand-real-sustainability-plan/. Accessed on June 28, 2018.

Hamerschlag, Kari, Anna Lappe, and Stacy Malkan. 2015. "Spinning Food: How Food Industry Front Groups and Covert Communications Are Shaping the Story of Food." *Friends of the Earth*. https://foe.org/resources/spinning-food-how-food-industry-front-groups-and-covert-communications-are-shaping-the-story-of-food/. Accessed on February 14, 2017.

Hansen, Arve. 2018. "Meat Consumption and Capitalist Development: The Meatification of Food Provision and Practice in Vietnam." *Geoforum* 93:57–68.

Harari, Yuval Noah. 2015. "Industrial Farming Is One of the Worst Crimes in History." *Guardian*, September 25, 2015. https://www.theguardian.com/books/2015/sep/25/industrial-farming-one-worst-crimes-history-ethical-question. Accessed on May 20, 2018.

Harbaugh, Rick. 1998. *Chinese Characters: A Genealogy and Dictionary*. New Haven, CT: Yale University Press.

Harrell, E. 2009. "Where's the Beef? Ghent Goes Vegetarian." *Time*, May 27. http://content.time.com/time/world/article/0,8599,1900958,00.html. Accessed on September 22, 2017.

Hartung, J. 2003. "Contribution of Animal Husbandry to Climatic Changes." In *Interactions between Climate and Animal Production*, ed. N. Lacetera, U. Bernabucci, H. H. Khalifa, B. Ronchi, and A. Nardone, 73–80. Wageningen, The Netherlands: Wageningen Academic Publishers.

Harvey, Celia, and William A. Haber. 1999. "Remnant Trees and the Conservation of Biodiversity in Costa Rican Pastures." *Agroforestry Systems* 44:37–68.

Haumann, B. 2017. "U.S. Organic Industry Continues to Grow." In *The World of Organic Agriculture: Statistics and Emerging Trends*, ed. H. Willer and J. Lernoud, 250–256. Bonn: Research Institute of Organic Agriculture, Frick and IFOAM–Organics International.

He, Meddy. 2013. "Cargill Inaugurates Integrated Poultry Operation in Lai'an, Anhui." https://www.cargill.com/news/releases/2013/NA3077688.jsp. Accessed on February 15, 2019.

Health Canada. 2016. "Health Canada and Canadian Food Inspection Agency Approve AquAdvantage Salmon." Statement issued May 19. Ottawa, ON: Health Canada. Government of Canada. https://www.canada.ca/en/health-canada/news/2016/05/health-canada-and-canadian-food-inspection-agency-approve-aquadvantage-salmon.html. Accessed on February 12, 2019.

Heffernan, William D. 1998. "Agriculture and Monopoly Capital." *Monthly Review* 50 (3): 46–59.

Heffernan, William D., Mary Hendrickson, and Robert Gronski. 1999. *Consolidation in the Food and Agriculture System*. Report to the National Farmers Union. Columbia: Department of Rural Sociology, University of Missouri.

Hinrichs, C. Clare, and Rick Welsh. 2003. "The Effects of the Industrialization of US Livestock Agriculture on Promoting Sustainable Production Practices." *Agriculture and Human Values* 20:125–141.

Hirsch, Veronica. 2003. *Detailed Discussions of Legal Protections of the Domestic Chicken in the United States and Europe*. Animal Legal & Historical Center, Michigan State University College of Law, East Lansing. http://www.animallaw.info/articles/dduschick.htm. Accessed on January 18, 2006.

Hirtzer, Michael. 2016. "Pork Giant Smithfield Skips Middlemen in Grain Supply Chain." *Reuters*, December 30. https://www.reuters.com/article/us-smithfield-foods-grains-analysis-idUSKBN14J0FA. Accessed on July 20, 2017.

Hodal, Kate. 2016. "Thailand: Poultry Workers Cry Fowl amid Claim They 'Slept on Floor Next to 28,000 Birds.'" *Guardian*, August 1. https://www.theguardian.com/global-development/2016/aug/01/thai-chicken-farm-workers-slept-on-the-floor-next-to-28000-birds. Accessed on August 4, 2016.

Horlings, L. G., and T. K. Marsden. 2011. "Towards the Real Green Revolution? Exploring the Conceptual Dimensions of a New Ecological Modernisation of Agriculture That Could 'Feed the World.'" *Global Environmental Change* 21 (2): 441–452.

Horowitz, Roger. 1997. *Negro and White, Unite and Fight!: A Social History of Industrial Unionism in Meatpacking, 1930–90*. Chicago: University of Illinois Press.

Howard, Philip H. 2009. "Visualizing Consolidation in the Seed Industry." *Sustainability* 1:1266–1287.

Howard, Philip H. 2016. *Concentration and Power in the Food System: Who Controls What We Eat?* London: Bloomsbury Academic.

References

Hsu, Vera Y. N., and Francis L. K. Hsu. 1977. "Modern China: North." In *Food in Chinese Culture: Anthropological and Historical Perspectives*, ed. K. C. Chang, 295–316. New Haven, CT: Yale University Press.

Hunter, Tera W. 1997. *To 'Joy My Freedom: Southern Black Women's Lives and Labors after the Civil War*. Cambridge, MA: Harvard University Press.

Hvistendahl, Mara. 2012. "China Takes Aim at Rampant Antibiotic Resistance." *Science* 336 (6083): 795.

Ilbery, Brian, and Damian Maye. 2005a. "Alternative (Shorter) Food Supply Chains and Specialist Livestock Products in the Scottish—English Borders." *Environment and Planning A* 37 (5): 823–844.

Ilbery, Brian, and Damian Maye. 2005b. "Food Supply Chains and Sustainability: Evidence from Specialist Food Producers in the Scottish/English Borders." *Land Use Policy* 22: 331–344.

ILRI (International Livestock Research Institute). 2018. "Why Livestock Matter." https://www.ilri.org/whylivestockmatter. Accessed on February 18, 2019.

Imhoff, Daniel, ed. 2010. *The CAFO Reader: The Tragedy of Industrial Animal Factories*. Berkeley: University of California Press.

Immigration and Refugee Board of Canada. 1999. *Ecuador: Role and Effectiveness of the IERAC in Conflicts with Indigenous People and Landowners; Role of Its Financial Directors, and Mistreatment of Its Officers by Conflicting Parties (1992–1999)*, February 1. ECU31219.E. http://www.refworld.org/docid/3ae6ac4c50.html. Accessed on February 2, 2018.

IPCC. 2006. *2006 IPCC Guidelines for National Greenhouse Gas Inventories*. Vol. 4: Agriculture, Forestry and Other Land Use. http://www.ipcc-nggip.iges.or.jp/public/2006gl/vol4.html. Accessed on January 2, 2019.

IPCC. 2013. "Summary for Policymakers." In *Climate Change 2013: The Physical Science Basis. Contribution of Working Group I to the Fifth Assessment Report of the Intergovernmental Panel on Climate Change*, ed. T. F. Stocker, D. Qin, G.-K. Plattner, M. Tignor, S. K. Allen, J. Boschung, A. Nauels, Y. Xia, V. Bex, and P. M. Midgley. Cambridge, UK, and New York: Cambridge University Press. http://www.ipcc.ch/pdf/assessment-report/ar5/wg1/WG1AR5_SPM_FINAL.pdf. Accessed on January 2, 2019.

IPES-Food. 2016. *From Uniformity to Diversity: A Paradigm Shift from Industrial Agriculture to Diversified Agroecological Systems*. International Panel of Experts on Sustainable Food Systems. http://www.ipes-food.org/images/Reports/UniformityToDiversity_FullReport.pdf. Accessed on February 2, 2019.

IPES-Food. 2017. *Too Big to Feed: Exploring the Impacts of Mega-Mergers, Consolidation, and Concentration of Power in the Food System*. International Panel of Experts on

Sustainable Food Systems. http://www.ipes-food.org/images/Reports/Concentration_FullReport.pdf.

Isaacs, Krista B., Sieglinde S. Snapp, James D. Kelly, and Kimberly R. Chung. 2016. "Farmer Knowledge Identifies a Competitive Bean Ideotype for Maize–Bean Intercrop Systems in Rwanda." *Agriculture & Food Security* 5 (1): 1–18.

Isakson, Johnny, and Christopher A. Coons. Letter from Senators Johnny Isakson and Christopher A. Coons to President Jacob Zuma, September 11, 2015. https://agoa.info/images/documents/5849/9-11114-joint-jhi-coons-letter-to-zuma-re-agoa-and-paris-agreement.pdf. Accessed on February 9. 2019.

Jakubek, Joseph, and Spencer D. Wood. 2018. "Emancipatory Empiricism: The Rural Sociology of W. E. B. Du Bois." *Sociology of Race and Ethnicity* 4 (1): 14–34.

James, Clive. 2014. *Global Status of Commercialized Biotech/ GM Crops: 2014*. Executive summary. International Service for the Acquisition of Agri-Biotech Applications (ISAAA). Brief No. 49. http://www.isaaa.org/resources/publications/briefs/49/. Accessed on February 2, 2019.

JBS. 2016. *Annual and Sustainability Report*. http://bit.ly/2f1PbeU. Accessed on February 2, 2019.

Ji, Xiuling, Qunhui Shen, Fang Liu, Jing Ma, Gang Xu, Yuanlong Wand, and Minghong Wu. 2012. "Antibiotic Resistance Gene Abundances Associated with Antibiotics and Heavy Metals in Animal Manures and Agricultural Soils Adjacent to Feedlots in Shanghai; China." *Journal of Hazardous Materials* 235–236: 178–185.

Kabel, Marcus. 2006. "Tyson Foods Sees Higher Meat Prices." *Washington Post*, November 13. http://www.washingtonpost.com/wp-dyn/content/article/2006/11/13/AR2006111300282.html. Accessed on March 9, 2014.

Kadim, I. T., O. Mahgoub, S. Baqir, B. Faye, and R. Purchas. 2015. "Cultured Meat from Muscle Stem Cells: A Review of Challenges and Prospects." *Journal of Integrative Agriculture* 14 (2): 222–233.

Kagawa, Masahiro, and Conner Bailey. 2006. "Trade Linkages in Shrimp Exports: Japan, Thailand, and Vietnam." *Journal of Development Studies* 24 (3): 303–319.

Kalkowski, John. 2018. "Top 50 Food Packaging Companies of 2018." *Packaging Strategies*, July 11. https://www.packagingstrategies.com/articles/90537-top-50-food-packaging-companies-of-2018. Accessed on July 12, 2018.

Kandel, William, and Emilio A. Parrado. 2005. "Restructuring of the US Meat Processing Industry and New Hispanic Migrant Destinations." *Population and Development Review* 31 (3): 447–471.

Kay, Steve. 2018. "Protein and Profits." *MEAT+POULTRY*, January. http://www.meatpoultry.com/Writers/Other-Contributors/Protein-and-profits.aspx. Accessed on February 2, 2018.

Kelly, Max. 2016. "Animals in International Development, Ethics, Dilemmas and Possibilities." *Ethical Issues in Poverty Alleviation* 14:113–132.

Kennedy, E. H., J. Johnston, and J. R. Parkins. 2017 "Small-p Politics: How Pleasurable, Convivial and Pragmatic Political Ideals Influence Engagement in Eat-Local Initiatives." *British Journal of Sociology* 69 (3): 670–690.

Keogh M., M. Henry, and N. Day. 2016. *Enhancing the Competitiveness of the Australian Livestock Export Industry*. Research report by the Australian Farm Institute, August. Surry Hills, Australia.

Kimura, Aya Hirata, and Mima Nishiyama. 2008. "The Chisan-Chisho Movement: Japanese Local Food Movement and Its Challenges." *Agriculture and Human Values* 25 (1): 49–64.

Kinchy, Abby. 2012. *Seeds, Science, and Struggle: The Global Politics of Transgenic Crops*. Cambridge, MA: MIT Press.

Kleinfield, N. R. 1984. "America Goes Chicken Crazy." *New York Times*, December 9. http://www.nytimes.com/1984/12/09/business/america-goes-chicken-crazy.html?pagewanted=all. Accessed on February 2, 2019.

Kloppenburg, J. 2004. *First the Seed: The Political Economy of Plant Biotechnology*. Madison: University of Wisconsin Press.

Kolbert, Elizabeth. 2014. *The Sixth Extinction: An Unnatural History*. New York: Henry Holt and Co.

Kowitt, Beth. 2017. "Bill Gates and Richard Branson Are Investing in This Clean Meat Startup." *Fortune*, August 23. http://fortune.com/2017/08/23/bill-gates-richard-branson-invest-meat/. Accessed on August 28, 2017.

Krkošek, M., M. A. Lewis, A. Morton, L. N. Frazer, and J. P. Volpe. 2006. "Epizootics of Wild Fish Induced by Farm Fish." *Proceedings of the National Academy of Sciences* 103 (42): 15506–15510.

La Via Campesina. 2009. "A G8 on Agriculture without Farmers = More Hunger and Poverty." Press release, April 21, 2009. https://viacampesina.org/en/a-g8-on-agriculture-without-farmers-more-hunger-and-poverty/. Accessed on September 22, 2017.

Lapegna, Pablo. 2016. *Soybeans and Power: Genetically Modified Crops, Environmental Politics, and Social Movements in Argentina*. New York: Oxford University Press.

Lappe, Frances Moore, Jennifer Clapp, Molly Anderson, Robin Broad, Ellen Messer, Thomas Pogge, and Timothy Wise. 2013. "How We Count Hunger Matters." *Ethics & International Affairs* 27 (3): 251–259.

Lassen, Jesper, Peter Sandøe, and Björn Forkman. 2006. "Happy Pigs Are Dirty!–Conflicting Perspectives on Animal Welfare." *Livestock Science* 103 (3): 221–230.

Lavers, Tom. 2012. "'Land Grab' as Development Strategy? The Political Economy of Agricultural Investment in Ethiopia." *Journal of Peasant Studies* 39 (1): 105–132.

Leahy, Eimear, Seán Lyon, and Richard S. J. Tol. 2010. "An Estimate of the Number of Vegetarians in the World." Working Paper No. 340. Dublin: Economic and Social Research Institute.

Leonard, Christopher. 2014. *The Meat Racket: The Secret Takeover of America's Food Business*. New York: Simon and Schuster.

Lerner, Amy, Thomas K. Rudel, Laura Schneider, Megan McGroddy, Diana Burbano, and Carlos Mena. 2015. "The Spontaneous Emergence of Silvo-Pastoral Landscapes in the Ecuadorian Amazon: Patterns and Processes." *Regional Environmental Change* 15 (7): 1421–1431.

Li, Jian. 2010. "The Decline of Household Pig Farming in Rural Southwest China: Socioeconomic Obstacles and Policy Implications." *Culture & Agriculture* 32 (2): 61–77.

Li, Peter J. 2009. "Exponential Growth, Animal Welfare, Environmental and Food Safety Impact: the Case of China's Livestock Production." *Journal of Agricultural and Environmental Ethics* 22 (3): 217–240.

Linebaugh, Peter. 2003. *The London Hanged: Crime and Civil Society in the Eighteenth Century*. London: Verso.

Little, Jo, Brian Ilbery, David Watts, Andrew Gilg, and Sue Simpson. 2012. "Regionalization and the Rescaling of Agro-Food Governance: Case Study Evidence from Two English Regions." *Political Geography* 31 (2): 83–93.

Liu, Jianguo, Vanessa Hull, Mateus Batistella, Ruth DeFries, Thomas Dietz, Feng Fu, Thomas Hertel, et al. 2013. "Framing Sustainability in a Telecoupled World." *Ecology and Society* 18 (2): 26.

Loconto, Allison, and Eve Fouilleux. 2014. "Politics of Private Regulation: ISEAL and the Shaping of Transnational Sustainability Governance." *Regulation & Governance* 8:166–185.

Luo, Yi, Daqing Mao, Michal Rysz, Qixing Zhou, Hongjie Zhang, Lin Xu, and Pedro J. J. Alvarez. 2010. "Trends in Antibiotic Resistance Genes Occurrence in the Haihe River, China." *Environmental Science and Technology* 44 (19): 7220–7225.

MacIntosh, Julie. 2011. *Dethroning the King: The Hostile Takeover of Anheuser-Busch, an American Icon*. Hoboken, NJ: John Wiley & Sons.

References

Macleod, M., Pierre Gerber, A. Mottet, G. Tempio, A. Falcucci, Carolyn Opio, Theun Vellinga, Benjamin Henderson, and Henning Steinfeld. 2013. *Greenhouse Gas Emissions from Pig and Chicken Supply Chains—A Global Life Cycle Assessment.* Food and Agriculture Organization of the United Nations. http://www.fao.org/docrep/018/i3460e/i3460e.pdf. Accessed on December 5, 2018.

Magalhaes, Luciana, and Rogerio Jelmayer. 2017. "Brazil Prosecutors' Request Freeze on JBS Chairman's Assets." *Wall Street Journal*, February 7, sec. Business. https://www.wsj.com/articles/brazil-prosecutors-request-freeze-on-jbs-chairmans-assets-1486489361. Accessed on February 14, 2017.

Maisonnave, Fabiano. 2017. "Troubled Meatpacker JBS Sanctioned over Amazon Deforestation." *Climate Change News*, March 31. http://www.climatechangenews.com/2017/03/31/troubled-meatpacker-jbs-sanctioned-amazon-deforestation. Accessed on July 17, 2018.

Manning, L., R. N. Baines, and S. A. Chadd. 2007. "Trends in the Global Poultry Meat Supply Chain." *British Food Journal* 109 (5): 332–342.

Mapes, Lynda V. 2017. "State Puts Hold on New Pens for Farmed Atlantic Salmon after Mass Escape in Puget Sound." *Seattle Times*, August 28. http://www.seattletimes.com/seattle-news/environment/state-puts-hold-on-new-pens-for-farmed-atlantic-salmon-after-mass-escape-in-puget-sound/. Accessed on February 12, 2019.

Marrow, Helen B. 2011. *New Destination Dreaming: Immigration, Race, and Legal Status in the Rural American South.* Stanford, CA: Stanford University Press.

Marx, Karl. (1867) 1992. *Capital: A Critique of Political Economy.* New York: Penguin Classics.

Masiga, W. N., and S. J. M. Munyua. 2005. "Global Perspectives on Animal Welfare: Africa." *Revue Scientifique Et Technique-Office International Des Epizooties* 24 (2): 579–589.

Maurer, Donna. 2002. *Vegetarianism: Movement or Moment?* Philadelphia: Temple University Press.

Mbuza, Francis, Rosine Manishimwe, Janvier Mahoro, Thomas Simbankabo, and Kizito Nishimwe. 2017. "Characterization of Broiler Poultry Production System in Rwanda." *Tropical Animal Health and Production* 49 (1): 71–77.

McCartin, Joseph A. 2011. *Collision Course: Ronald Reagan, the Air Traffic Controllers, and the Strike That Changed America.* Oxford, UK: Oxford University Press.

McDougal, Tony. 2018. "China Hits Brazil's Poultry with Anti-Dumping Deposits." *Poultry World*, June 11. https://www.poultryworld.net/Meat/Articles/2018/6/China-hits-Brazils-poultry-with-anti-dumping-deposits-295216E/?cmpid=NLC%7C worldpoultry%7C2018-06-13%7CChina_hits_Brazil?s_poultry_with_anti-dumping_deposits. Accessed on June 12, 2018.

McMichael, Philip D. 2009a. "The World Food Crisis in Historical Perspective." *Monthly Review* 61 (3): n.p. https://monthlyreview.org/2009/07/01/the-world-food-crisis-in-historical-perspective/. Accessed on December 5, 2018.

McMichael, Philip D. 2009b. "A Food Regime Genealogy." *Journal of Peasant Studies* 36 (1): 139–169.

McMichael, Philip D. 2017. *Development and Social Change: A Global Perspective*. 6th ed. London: SAGE Publications.

Micheletti, M., and D. Stolle 2012. "Vegetarianism—A Lifestyle Politics?" In *Creative Participation: Responsibility-Taking in the Political World*, ed. Michele Micheletti and Andrew S. McFarland, 127–147. New York: Paradigm Publishers.

Ministry of Agriculture and Animal Resources (MINAGRI). 2012. *Strategy and Investment Plan to Strengthen the Poultry Industry in Rwanda*. Rwanda: MINAGRI.

Ministry of Agriculture of the People's Republic of China. 2009. *China Agricultural Development Report*. Beijing: China Agricultural Press.

Mize, Ronald L., and Alicia C. S. Swords. 2010. *Consuming Mexican Labor: From the Bracero Program to NAFTA*. Toronto: University of Toronto Press.

Monday Campaign. 2017. "The Movement." *Meatless Monday Global*. http://www.meatlessmonday.com/the-global-movement/. Accessed on September 22, 2017.

Morris, Carol, and James Kirwan. 2006. "Vegetarians: Uninvited, Uncomfortable or Special Guests at the Table of the Alternative Food Economy?" *Sociologia Ruralis* 46 (3): 192–213.

Mulambu, J., M. Andersson, M. Palenberg, W. Pfeiffer, A. Saltzman, E. Birol, A. Oparinde, E. Boy, D. Asare-Marfo, A. Lubobo, and C. Mukankusi. 2017. "Iron Beans in Rwanda: Crop Development and Delivery Experience." *African Journal of Food, Agriculture, Nutrition and Development* 17 (2): 12026–12050.

Murenzi, Romain, and Mike Hughes. 2006. "Building a Prosperous Global Knowledge Economy in Africa: Rwanda as a Case Study." *International Journal of Technology and Globalization* 2 (3–4): 252–267.

Musolin, Kristin, Jessica G. Ramsey, James T. Wassell, David L. Hard, and Charles Mueller. 2014. *Evaluation of Musculoskeletal Disorders and Traumatic Injuries among Employees at a Poultry Processing Plant*. Washington, DC: USDHHS, CDC, NIOSH. https://www.cdc.gov/niosh/hhe/reports/pdfs/2012-0125-3204.pdf. Accessed on December 5, 2018.

Nam, Ki-Chang, Cheorun Jo, and Mooha Lee. 2010. "Meat Products and Consumption Culture in the East." *Meat Science* 86 (1): 95–102.

National Bureau of Statistics. n.d. *National Data*. Beijing, China: National Bureau of Statistics of China. http://data.stats.gov.cn/english/index.htm. Accessed on December 2, 2018.

Naylor, R. L., R. J. Goldburg, J. H. Primavera, N. Kautsky, M. C. Beveridge, J. Clay, C. Folke, J. Lubchenco, H. Mooney, and M. Troell. 2000. "Effect of Aquaculture on World Fish Supplies." *Nature* 405 (6790): 1017–1024.

Naylor, Rosamond, Henning Steinfeld, Walter Falcon, James Galloway, Vaclav Smil, Eric Bradford, Jackie Alder, and Harold Mooney. 2005. "Losing the Links between Livestock and Land. *Science* 310 (5754): 1621–1622.

Naylor, Rosamond L., Ronald W. Hardy, Dominique P. Bureau, Alice Chiu, Matthew Elliott, Anthony P. Farrell, Ian Forster, Delbert M. Gatlin, Rebecca J. Goldburg, Katheline Hua, and Peter D. Nichols. 2009. "Feeding Aquaculture in an Era of Finite Resources." *Proceedings of the National Academy of Sciences* 106 (2): 15103–15110.

NCC (National Chicken Council). 2013. "History of the National Chicken Council." National Chicken Council. https://www.nationalchickencouncil.org/about-ncc/history/. Accessed on March 2, 2018.

NCC (National Chicken Council). 2018. "Per Capita Consumption of Poultry and Livestock, 1965 to Estimated 2018, in Pounds." National Chicken Council. https://www.nationalchickencouncil.org/about-the-industry/statistics/per-capita-consumption-of-poultry-and-livestock-1965-to-estimated-2012-in-pounds/. Accessed on March 2, 2018.

NCD Risk Factor Collaboration. 2016. "Trends in Adult Body-Mass Index in 200 Countries from 1975 to 2014: A Pooled Analysis of 1698 Population-Based Measurement Studies with 19.2 Million Participants." *Lancet* 387: 1377–1396.

Neo, Harvey and Jody Emel. 2017. *Geographies of Meat: Politics, Economy and Culture*. New York: Routledge.

Nepstad, Daniel C., Claudia Stickler, and Oriana Almeida. 2006. "Globalization of the Amazon Soy and Beef Industries: Opportunities for Conservation." *Conservation Biology* 20:1595–1603.

Netting, Robert McC. 1993. *Smallholders, Householders: The Ecology of Small Scale, Sustainable Agriculture*. Stanford, CA: Stanford University Press.

Newport, F. 2012. "In U.S., 5% Consider Themselves Vegetarians." *Gallup*. http://www.gallup.com/poll/156215/consider-themselves-vegetarians.aspx. Accessed on July 10, 2018.

Nitzan, Jonathan, and Shimshon Bichler. 2009. *Capital as Power: A Study of Order and Creorder*. New York: Routledge.

Nyéléni International Steering Committee. 2007. *Synthesis Report*. Nyéléni International Steering Committee, March. https://nyeleni.org/IMG/pdf/31Mar2007NyeleniSynthesisReport-en.pdf. Accessed on September 22, 2019.

O'Connor, James. 1973. *Fiscal Crisis of the State*. New York: St. Martin's Press.

Obeysekare, Eric, Khanjan Mehta, and Carleen Maitland. 2017. "Defining Success in a Developing Country's Innovation Ecosystem: The Case of Rwanda." Paper presented at the Global Humanitarian Technology Conference (GHTC), October 19–22, San Jose, CA.

Odem, Mary E. 2009. "Latino Immigrants and the Politics of Space in Atlanta." In *Latino Immigrants and the Transformation of the U.S. South*, ed. M. E. Odem and E. C. Lacy, 112–125. Athens: University of Georgia Press.

Odem, Mary E., and Elaine Cantrell Lacy. 2009. *Latino Immigrants and the Transformation of the U.S. South*. Athens: University of Georgia Press.

OECD/FAO. 2014. *OECD-FAO Agricultural Outlook 2014*. Organization for Economic Cooperation and Development/Food and Agriculture Organization of the United Nations. Rome: OECD Publishing. https://doi.org/10.1787/agr_outlook-2014-en.

OECD/FAO. 2016. *OECD-FAO Agricultural Outlook 2016–2025*. Organization for Economic Cooperation and Development/Food and Agriculture Organization of the United Nations. Paris: OECD Publishing.

OIE (World Organization for Animal Health). 2011. *Animal Welfare in OIE Member Countries and Territories in the SADC Region: Summaries of Baseline Country Assessments*. World Organization for Animal Health, Sub-regional Representation for Southern Africa, Gaborone. Paris: OIE Publishing.

Oliveira, Gustavo de L. T. 2015. The Geopolitics of Brazilian Soybeans. *Journal of Peasant Studies* 43 (2): 348–372.

Oliveira, Gustavo de L. T., and Mindi Schneider. 2016. "The Politics of Flexing Soybeans: China, Brazil and Global Agroindustrial Restructuring." *Journal of Peasant Studies* 43 (1): 167–194.

Oliveira, Gustavo de L. T., and Susana Hecht. 2016. "Sacred Groves, Sacrifice Zones and Soy Production: Globalization, Intensification and Neo-Nature in South America." *Journal of Peasant Studies* 43 (2): 251–285.

Olivier, J. G. J., K. M. Schure, and J. A. H. W. Peters. 2017. *Trends in Global CO_2 and Total Greenhouse Gas Emissions: 2017 Report*. The Hague. http://www.pbl.nl/sites/default/files/cms/publicaties/pbl-2017-trends-in-global-co2-and-total-greenhouse-gas-emissons-2017-report_2674.pdf. Accessed on February 2, 2019.

Österblom, Henrik, Jean-Baptiste Jouffray, Carl Folke, Beatrice Crona, Max Troell, Andrew Merrie, and Johan Rockström. 2015. "Transnational Corporations as 'Keystone Actors' in Marine Ecosystems." *PLOS/One* 10 (5): e0127533. https://doi.org/10.1371/journal.pone.0127533. Accessed on February 15, 2019.

Otero, Gerardo. 2018. *The Neoliberal Diet: Healthy Profits, Unhealthy People*. Austin: University of Texas Press.

Otero, Gerardo, Efe Can Gürcan, Gabriela Pechlaner, and Giselle Liberman. 2018. "Food Security, Obesity, and Inequality: Measuring the Risk of Exposure to the Neoliberal Diet." *Journal of Agrarian Change* 18:536–554.

Oxfam America. 2016. "Lives on the Line: The Human Cost of Cheap Chicken." Oxfam Poultry Workers Brief, October 26. https://www.oxfamamerica.org/static/media/files/Oxfam_Poultry_Workers_Brief_October_26_5UhpYvF.pdf. Accessed on January 10, 2019.

Pachirat, T. 2013. *Every Twelve Seconds: Industrialized Slaughter and the Politics of Sight.* New Haven, CT: Yale University Press.

Patel, Raj. 2016. "How Society Subsidizes Big Food and Poor Health." *JAMA Internal Medicine* 176 (8): 1132–1133.

Patel, Raj, and Jason W. Moore. 2018. *History of the World in Seven Cheap Things: A Guide to Capitalism, Nature, and the Future of the Planet.* Berkeley: University of California Press.

Peine, Emily. 2013. "Trading on Pork and Beans: Agribusiness and the Construction of the Brazil-China-Soy-Pork Commodity Complex." In *The Ethics and Economics of Agrifood Competition*, ed. Harvey S. James Jr., 193–210. New York: Springer.

People's Daily. 2007. "China's Fourth Baby Boom." *People's Daily*, April 18. http://en.people.cn/200704/18/eng20070418_367611.html. Accessed on May 13, 2018.

Perez, Karni R. 2006. *Fishing for Gold: The Story of Alabama's Catfish Industry.* Tuscaloosa: University of Alabama Press.

Perrow, Charles. 1984. *Normal Accidents: Living with High Risk Systems.* New York: Basic Books.

Philpott, Tom. 2013. "Are We Becoming China's Factory Farm?" *Mother Jones*, December. http://www.motherjones.com/media/2014/03/china-factory-farm-america-pork. Accessed on February 15, 2017.

Philpott, Tom. 2015. "Bird Flu Is Slamming Factory Farms but Sparing Backyard Flocks. Why?" *Mother Jones*, May 20. http://www.motherjones.com/tom-philpott/2015/05/ongoing-bird-flu-crisis-stumps-experts. Accessed on February 14, 2017.

Pi, Chendong. 2014. *Fair or Fowl? Industrialization of Poultry Production in China.* Minneapolis, MN: Institute for Agriculture and Trade Policy.

Pigatto, Gessuir, and Guiliana Aparecida Santini Pigatto. 2015. "The Strategy for Internationalization of Brazilian Meat Industries and the Role of the Development Bank." *Informe GEPEC* 19 (2): 126–146.

Pilgrim's Pride Corporation. 2008. "Pilgrim's Pride Issues Statement in Response to U.S. Department of Homeland Security's Immigration and Customs

Enforcement Action at Five Company Facilities." *SecurityInfo*, April 16. https://www
.securityinfowatch.com/article/10545681/pilgrims-pride-issues-statement-in
-response-to-us-department-of-homeland-securitys-immigration-and-customs
-enforcement-actio. Accessed on February 15, 2019.

Pollan, Michael. 2006. *The Omnivore's Dilemma: A Natural History of Four Meals*. New York: Penguin Press

Ponte, S. 2014. "'Roundtabling' Sustainability: Lessons from the Biofuel Industry." *Geoforum* 54: 261–271.

Popkin, Barry M. 2002. "The Shift in Stages of the Nutrition Transition in the Developing World Differs from Past Experiences!" *Public Health Nutrition* 5 (1): 205–214.

PRB. 2010. *Despite Wide-Ranging Benefits, Girls' Education and Empowerment Overlooked in Developing Countries*. Population Reference Bureau (PRB), last modified April 27, 2010. https://www.prb.org/girlseducation/. Accessed on February 18, 2019.

Public Citizen. 2004. *Smithfield Foods: A Corporate Profile*. Washington, DC: Public Citizen. http://www.citizen.org/documents/Smithfield.pdf. Accessed on September 13, 2014.

Qiao, Fangbin, Jikun Huang, Dan Wang, Huaiju Liu, and Bryan Lohmar. 2016. "China's Hog Production: From Backyard to Large-Scale." *China Economic Review* 38 (2016): 199–208.

Rae, Allan. 2008. "China's Agriculture, Smallholders and Trade: Driven by the Livestock Revolution?" *Australian Journal of Agricultural and Resource Economics* 52 (3): 283–302.

Ransom, Elizabeth. 2007. "The Rise of Agricultural Animal Welfare Standards as Understood through a Neo-Institutional Lens." *International Journal of Sociology of Agriculture and Food* 15 (3): 26–44.

Rao, Idupulapati, Michael Peters, Aracely Castro, Rainer Schultze-Kraft, Douglas White, Miles Fisher, John Miles, et al. 2015. "LivestockPlus—The Sustainable Intensification of Forage-Based Agricultural Systems to Improve Livelihoods and Ecosystem Services in the Tropics." *Tropical Grasslands—Forrajes Tropicales* 3:59–82.

Raper, Arthur Franklin. (1936) 2005. *Preface to Peasantry: A Tale of Two Black Belt Counties*. Columbia: University of South Carolina Press.

Ready, Valerie. 1978. "Gold Kist Workers Walk Out; Salaries, Conditions, at Issue." *The Athens Daily News*, April.

Redford, Kent. 1991. "The Ecological Noble Savage." *Cultural Survival Quarterly* 15 (1): 46–48.

Regan, Tom. 1983. "Animal Rights, Human Wrongs." In *Ethics and Animals*, ed. H. B. Miller and W. H. Williams, 19–43. New York: Humana Press.

Reyna v. ConAgra Foods, Inc. 2007. Civil Action No. 3:04:cv-39 (CDL). U.S. District Court for the Middle District of Georgia. 506 F. Supp. 2d 1363.

Ribas, Vanesa. 2015. *On the Line: Slaughterhouse Lives and the Making of the New South.* Oakland: University of California Press.

Richardson, Robin Y. 2011. "Former Pilgrim's President Details Founder's Practices." *Marshall News Messenger*, June 21. http://www.marshallnewsmessenger.com/news/2011/jun/21/former-pilgrims-president-details-founders-practic/. Accessed on February 14, 2017.

Robertson, G. Philip, Tom W. Bruulsema, Ron J. Gehl, David Kanter, Denise L. Mauzerall, C. Alan Rotz, and Candiss O. Williams. 2013. "Nitrogen–Climate Interactions in US Agriculture." *Biogeochemistry* 114 (1–3): 41–70.

Robinson, Brian E., Margaret B. Holland, and Lisa Naughton-Treves. 2014. "Does Secure Land Tenure Save Forests? A Meta-Analysis of the Relationship between Land Tenure and Tropical Deforestation." *Global Environmental Change* 29: 281–293.

Rosegrant, M. W., M. Fernandez, A. Sinha, J. Alder, Helal Ahammad, Charlotte de Fraiture, B. Eickhour, et al. 2009. "Looking into the Future for Agriculture and AKST." In *International Assessment of Agricultural Knowledge, Science and Technology for Development (IAASTD): Agriculture at a Crossroads*, ed. B. D McIntyre, H. R. Herren, J. Wakhungu, and R. T. Watson, 307–376. Washington, DC: Island Press.

Rosendal, G. K, I. Olesen, and M. W. Tvedt. 2013. "Evolving Legal Regimes, Market Structures and Biology Affecting Access to and Protection of Aquaculture Genetic Resources." *Aquaculture* 402–403:97–105.

Rosenzweig, C., A. Iglesias, X. B. Yang, P. R. Epstein, and E. Chivian 2001. "Climate Change and Extreme Weather Events: Implications for Food Production, Plant Diseases, and Pests." *Global Change & Human Health* 2 (2): 90–104.

Rottenberg, Carmen. 2018. USDA FSIS Response to NCC, January 29. Washington, DC. https://www.fsis.usda.gov/wps/wcm/connect/235092cf-e3c0-4285-9560-e60cf6956df8/17-05-FSIS-Response-Letter-01292018.pdf?MOD=AJPERES. Accessed on January 5, 2019.

Rowland, Michael Pellman. 2017. "Tyson Foods Injects More Money into Plant-Based Meat." *Forbes*, December 10. https://www.forbes.com/sites/michaelpellmanrowland/2017/12/10/tyson-foods-plant-based-meat/#7fb0be267efa. Accessed on September 22, 2018.

Rudel, Thomas K., and Bruce Horowitz. 1993. *Tropical Deforestation: Small Farmers and Land Clearing in the Ecuadorian Amazon.* New York: Columbia University Press.

Rudel, Thomas K., Tuntiak Katan, and Bruce Horowitz. 2013. "Amerindian Livelihoods, Outside Interventions, and Poverty Traps in the Ecuadorian Amazon." *Rural Sociology* 78 (2): 167–185.

Rueda, Ximena, and Eric Lambin. 2013. "Linking Globalization to Local Land Uses: How Eco-Consumers and Gourmands Are Changing Colombian Coffee Landscapes." *World Development* 41:286–301.

Rulli, Maria Cristina, Antonio Saviori, and Paolo D'Odorico. 2013. "Global Land and Water Grabbing." *Proceedings of the National Academy of Sciences* 110 (3): 892–897.

Rwanda Development Board. 2018. "Agriculture." http://rdb.rw/demo2/investment-opportunities/agriculture/. Accessed in February 2018.

Sahota, A. 2016. "The Global Market for Organic Food and Drink." In *The World of Organic Agriculture: Statistics and Emerging Trends 2016*, ed. H. Willer and J. Lernoud, 134–137. Bonn: Research Institute of Organic Agriculture (FiBL), Frick and IFOAM—Organics, International.

Samenow, Jason. 2018. "Red-Hot Planet: All-Time Heat Records Have Been Set All over the World during the Past Week." *Washington Post*, July 5. https://www.washingtonpost.com/news/capital-weather-gang/wp/2018/07/03/hot-planet-all-time-heat-records-have-been-set-all-over-the-world-in-last-week/?utm_term=.f0dcb8d330a6. Accessed on August 10, 2018.

Sauer, Sérgio, and Sergio Pereira Leite. 2012. "Agrarian Structure, Foreign Investment in Land, and Land Prices in Brazil." *Journal of Peasant Studies* 39 (3–4): 873–898.

Schlosser, Eric. 2001. *Fast Food Nation: The Dark Side of the All-American Meal*. New York: Houghton Mifflin Harcourt.

Schlozman, Kay Lehman, Sidney Verba, and Henry E. Brady. 2012. *The Unheavenly Chorus: Unequal Political Voice and the Broken Promise of American Democracy*. Princeton, NJ: Princeton University Press.

Schmalzer, Sigrid. 2016. *Red Revolution, Green Revolution: Scientific Farming in Socialist China*. Chicago: The University of Chicago Press.

Schmidt, Blake. 2014. "Brazilian Barons Become Five Slaughterhouse Billionaires." *Bloomberg*, December 15. https://www.bloomberg.com/news/articles/2014-12-15/brazilian-barons-become-five-slaughterhouse-billionaires. Accessed on February 14, 2017.

Schneider, Mindi. 2011. *Feeding China's Pigs: Implications for the Environment, China's Smallholder Farmers and Food Security*. Minneapolis, MN: Institute for Agriculture and Trade Policy.

Schneider, Mindi. 2014. "Developing the Meat Grab." *The Journal of Peasant Studies* 41 (4): 613–633. https://doi.org/10.1080/03066150.2014.918959.

Schneider, Mindi. 2015. "What, Then, Is a Chinese Peasant? Nongmin Discourses and Agroindustrialization in Contemporary China." *Agriculture and Human Values* 32 (2): 331–346.

Schneider, Mindi. 2017a. "Dragon Head Enterprises and the State of Agribusiness in China." *Journal of Agrarian Change* 17 (1): 3–21.

Schneider, Mindi. 2017b. "Wasting the Rural: Meat, Manure and Agro-Industrialization in Contemporary China." *Geoforum* 78:89–97.

Schneider, Mindi, and Shefali Sharma. 2014. *China's Pork Miracle? Agribusiness and Development in China's Pork Industry.* Minneapolis, MN: Institute for Agriculture and Trade Policy.

Schouten, G., P. Leroy, and P. Glasbergen. 2012. "On the Deliberative Capacity of Private Multi-Stakeholder Governance: The Roundtables on Responsible Soy and Sustainable Palm Oil." *Ecological Economics* 83:42–50.

Schupp, J. L. 2016. "Just Where Does Local Food Live? Assessing Farmers' Markets in the United States." *Agriculture and Human Values* 33 (4): 827–841. doi:10.1007/s10460-015-9667-y. Accessed on November 10, 2018.

Schwartzman, Kathleen Crowley. 2013. *The Chicken Trail: Following Workers, Migrants, and Corporations across the Americas.* Ithaca, NY: Cornell University, ILR Press.

Seeth, Avantika. 2017. "Poultry Industry to Lose 50 000 Jobs." *City Press*, January 10. https://city-press.news24.com/Business/poultry-industry-to-lose-50-000-jobs-20170110. Accessed on January 29, 2019.

Seidman, G. 1994. *Manufacturing Militance: Workers' Movements in Brazil and South Africa, 1970–1985.* Berkeley: University of California Press.

Sharma, Shefali, and Sergio Schlesinger. 2017. *The Rise of Big Meat: Brazil's Extractive Industry.* Geneva: Institute for Agriculture and Trade Policy Europe.

Shi, Yan, Cunwang Cheng, Peng Lei, and Caroline Merrifield. 2011. "Safe Food, Green Food, Good Food: Chinese Community Supported Agriculture and the Rising Middle Class." *International Journal of Agricultural Sustainability* 9 (4): 551–558.

Shurtleff, William, H. T. Huang, and Akiko Aoyagi. 2014. *History of Soybeans and Soyfoods in China and Taiwan, and in Chinese Cookbooks, Restaurants, and Chinese Work with Soyfoods Outside China (1024 BCE to 2014): Extensively Annotated Bibliography and Sourcebook.* Lafayette, CA: Soyinfo Center.

Simon, Bryant. 2017. "The Geography of Silence: Food and Tragedy in Globalizing America." In *Food, Power, and Agency*, ed. Jürgen Martschukat and Bryant Simon, 83–102. New York: Bloomsbury Press.

Singer, Peter. 1975. *Animal Liberation: A New Ethic for Our Treatment of Animals.* New York: Avon.

Singer, R. 2017. "Neoliberal Backgrounding, the Meatless Monday Campaign, and the Rhetorical Intersections of Food, Nature, and Cultural Identity." *Communication, Culture & Critique* 10 (2): 344–364.

Sito, Peggy. 2016. "Chinese Pork Giant WH Group May Eye Major Acquisitions by End of 2017." *South China Morning Post*, November 20. http://www.scmp.com/business/companies/article/204771,9/chinese-pork-giant-wh-group-will-eye-major-acquisitions-end-2017. Accessed on February 14, 2017.

Smil, Vaclav. 2001. *Feeding the World: A Challenge for the Twenty-first Century*. Cambridge, MA: MIT Press.

Smith, Elliott Blair. 2014. "How Smithfield Foods Larded Its CEO Pay Package." *The Fiscal Times*, April 30. http://www.thefiscaltimes.com/Articles/2014/04/30/How-Pork-Giant-Smithfield-Foods-Larded-Its-CEO-s-Pay-Package. Accessed on July 16, 2018.

Smith, Jimmy. 2016. "Veganism Is Not the Key to Sustainable Development—Natural Resources Are Vital." *Guardian*, August 16. https://www.theguardian.com/global-development/2016/aug/16/veganism-not-key-sustainable-development-natural-resources-jimmy-smith. Accessed on February 18, 2019.

Smothers, Ronald. 1996. "Unions Try to Push Past Workers' Fears to Sign Up Poultry Plants in South." *New York Times*, January 30. http://www.nytimes.com/1996/01/30/us/unions-try-to-push-past-workers-fears-to-sign-up-poultry-plants-insouth.html. Accessed on February 15, 2019.

Song, Yang, Xuemei Li, and Lishi Zhang. 2014. "Food Safety Issues in China." *Iranian Journal of Public Health* 43 (9): 1299–1300.

Soussana, J. F., T. Tallec, and V. Blanfort. 2010. "Mitigating the Greenhouse Gas Balance of Ruminant Production Systems through Carbon Sequestration in Grasslands." *Animal* 4 (3): 334–350.

Starmer, Elanor, and Timothy A. Wise. 2007. "Living High on the Hog: Factory Farms, Federal Policy, and the Structural Transformation of Swine Production." Working Paper 07-04. Medford, MA: Global Development and Environment Institute, Tufts University.

Starmer, Elanor, Aimee Witteman, and Timothy A. Wise. 2006. "Feeding the Factory Farm: Implicit Subsidies to the Broiler Chicken Industry." Working Paper 06-03. Medford, MA: Global Development and Environment Institute, Tufts University.

Starr, A. 2010. "Local Food: A Social Movement?" *Cultural Studies ↔ Critical Methodologies* 10 (6): 479–490.

Steinfeld, Henning, Pierre Gerber, Tom Wassenaar, Vincent Castel, Mauricio Rosales, and Cees de Haan. 2006. *Livestock's Long Shadow: Environmental Issues and Options*. Rome: FAO. http://www.fao.org/docrep/010/a0701e/a0701e00.HTM.

Striffler, Steve. 2002. "Inside a Poultry Processing Plant: An Ethnographic Portrait." *Labor History* 43 (3): 305–313. https://doi.org/10.1080/0023656022000001797. Accessed on February 10, 2019.

Striffler, Steve. 2005. *Chicken: The Dangerous Transformation of America's Favorite Food.* New Haven, CT: Yale University Press.

Strom, Stephanie. 2016. "Tyson Foods, a Meat Leader, Invests in Protein Alternatives." *New York Times.* October 10. https://www.nytimes.com/2016/10/11/business/tyson-foods-a-meat-leader-invests-in-protein-alternatives.html. Accessed on August 10, 2018.

Stuesse, Angela. 2016. *Scratching Out a Living: Latinos, Race, and Work in the Deep South.* Berkeley: University of California Press.

Stull, Donald, and Michael J. Broadway. 2004. *Slaughterhouse Blues: The Meat and Poultry Industry in North America.* Belmont, CA: Wadsworth Publishing.

Tacon, Albert G. J., and Marc Metian. 2009. "Fishing for Aquaculture: Non-Food Use of Small Pelagic Forage Fish—A Global Perspective." *Reviews in Fisheries Science* 17:305–317.

Tao, Hongjun, and Chaoping Xie. 2015. "A Case Study of Shuanghui International's Strategic Acquisition of Smithfield Foods." *International Food and Agribusiness Management Review* 18 (1): 145–166.

Thomas N., R. Lucas, P. Bunting, A. Hardy, A. Rosenqvist, and M. Simard. 2017. "Distribution and Drivers of Global Mangrove Forest Change, 1996–2010." *PLoS ONE* 12 (6): e0179302. https://doi.org/10.1371/journal.pone.0179302. Accessed on February 12, 2019.

Thompson, E. P. 1967. "Time, Work-Discipline, and Industrial Capitalism." *Past & Present* 38 (1): 56–97.

Thornton, Philip K. 2010. "Livestock Production: Recent Trends, Future Prospects." *Philosophical Transactions of the Royal Society of London B: Biological Sciences* 365 (1554): 2853–2867.

Thorstad, Eva B., Ian A. Fleming, Philip McGinnity, Doris Soto, Vidar Wennevik, and Fred Whoriskey. 2008. *Incidence and Impacts of Escaped Farmed Atlantic Salmon Salmo salar in Nature.* Norwegian Institute for Nature Research (NINA). NINA Special Report 36. http://www.fao.org/3/a-aj272e.pdf. Accessed on January 20, 2019.

Towie, Narelle. 2014. "More Than 4000 Sheep Perish on Live Export Ship." *Sydney Morning Herald*, January 16. https://www.smh.com.au/environment/conservation/more-than-4000-sheep-perish-on-live-export-ship-20140116-30wf2.html. Accessed on August 10, 2018.

Tran, Nhuong, Conner Bailey, Norbert Wilson, and Michael Phillips. 2013. "Governance of Global Value Chains in Response to Food Safety and Certification Standards: The Case of Shrimp from Vietnam." *World Development* 45:325–336.

Troitino, Christina. 2018a. "FDA Gives Green Light to Impossible Foods' Bleeding Burgers." *Forbes*. July 27. https://www.forbes.com/sites/christinatroitino/2018/07/27/fda-gives-green-light-to-impossible-foods-bleeding-burgers/#2b48f83f6c47. Accessed on July 28, 2018.

Troitino, Christina. 2018b. "Missouri Becomes First State to Start Regulating Meat Alternative Labels." *Forbes*. August 31. https://www.forbes.com/sites/christinatroitino/2018/08/31/missouri-now-regulating-meat-alternative-labels-as-regulatory-war-gets-bloody/#69f18b168869. Accessed on September 8, 2018.

Tuomisto, H., and M. J. T. de Mattos. 2011. "Environmental Impacts of Cultured Meat Production." *Environmental Science & Technology* 45 (14): 6117-6123.

Turzi, Mariano. 2011. "The Soybean Republic." *Yale Journal of International Affairs* 6 (2): 59–68. http://yalejournal.org/pastissues_post/volume-6-issue-2-springsummer-2011/.

Turzi, Mariano. 2017. *The Political Economy of Agricultural Booms: Managing Soybean Production in Argentina, Brazil, and Paraguay*. Cham, Switzerland: Palgrave MacMillan.

Tyson Foods. n.d. "Investors." https://www.tysonfoods.com/investors. Accessed on February 14, 2017.

Tyson Foods. 2016. "Facts about Tyson Foods." https://web.archive.org/web/20181002124001/http://ir.tyson.com/investor-relations/investor-overview/tyson-factbook/default.aspx. Accessed on February 14, 2017.

Tyson Foods. 2018. "Around the World." https://www.tysonfoods.com/innovation/protein-leader/around-world. Accessed on December 18, 2018.

UGA CAES. 2017. *Poultry: The Largest Segment of Georgia Agriculture!* University of Georgia College of Agricultural and Environmental Sciences (CAES). http://www.caes.uga.edu/content/dam/caes-website/departments/poultry-science/documents/2017_Georgia_Poultry_Facts.pdf. Accessed on February 10, 2019.

Undercurrent News. 2016. "World's 100 Largest Seafood Companies 2016." *Undercurrent News*. https://www.undercurrentnews.com/report/undercurrent-news-worlds-100-largest-seafood-companies-2016/. Accessed on February 12, 2019.

United States Department of Agriculture (USDA) Foreign Agricultural Service (FAS). 2017a. *Livestock and Poultry: World Markets and Trade (April)*. Washington, DC: USDA. https://www.fas.usda.gov/data/livestock-and-poultry-world-markets-and-trade. Accessed on February 10, 2019.

United States Department of Agriculture (USDA) Foreign Agricultural Service (FAS). 2017b. *Oilseeds: World Markets and Trade (August)*. Washington, DC: USDA. https://www.fas.usda.gov/data/oilseeds-world-markets-and-trade. Accessed on December 4, 2018.

United States Department of Agriculture (USDA) Foreign Agricultural Service (FAS). n.d. *Production, Supply and Distribution* (PS&D) database. https://apps.fas.usda.gov/psdonline/app/index.html#/app/home. Accessed on December 4, 2018.

U.S. FDA (United States Food and Drug Administration). 2015. "FDA Has Determined That the AquAdvantage Salmon Is as Safe to Eat as Non-GE Salmon." *Consumer Update*, November 19. Washington, DC: U.S. Food & Drug Administration. https://www.fda.gov/ForConsumers/ConsumerUpdates/ucm472487.htm. Accessed on December 10, 2018.

U.S. GAO (United States Government Accountability Office). 2017. *Workplace Safety and Better Outreach, Collaboration, and Information Needed to Help Protect Workers at Meat and Poultry Plants*. GAO Report to Congressional Requesters, November.

U.S. House. 1951. J. D. Jewell Co. *Hearing before the Subcommittee on Labor and Labor-management Relations*. 82nd Congress, 1st Session. Washington, DC: Government Printing Office.

Van Hoyweghen, Saskia. 1999. "The Urgency of Land and Agrarian Reform in Rwanda." *African Affairs* 98 (392): 353–372.

van Huis, Arnold, Jan Van Itterbeeck, Harmke Klunder, Esther Mertens, Afton Halloran, Giulia Muir, and Paul Vantomme. 2013. *Edible Insects: Future Prospects for Food and Feed Security*. Rome: FAO.

Verbeke, W., A. Marcu, P. Rutsaert, R. Gaspar, B. Seibt, D. Fletcher, and J. Barnett. 2015. "'Would You Eat Cultured Meat?': Consumers' Reactions and Attitude Formation in Belgium, Portugal and the United Kingdom." *Meat Science* 102:49–58.

Vidal, John. 2010. "10 Ways Vegetarianism Can Help Save the Planet." *Guardian*, July 17. https://www.theguardian.com/lifeandstyle/2010/jul/18/vegetarianism-save-planet-environment. Accessed on May 20, 2018.

Wallace, Robert G. 2017. "Industrial Production of Poultry Gives Rise to Deadly Strains of Bird Flu H5Nx." *Independent Science News*, January. https://www.independentsciencenews.org/health/industrial-production-of-poultry-gives-rise-to-deadly-strains-of-bird-flu-h5nx/. Accessed on February 14, 2017.

Wang, Jimin, and Mariko Watanabe. 2008. *Pork Production in China: A Survey and Analysis of the Industry at a Lewis Turning Point*. Institute of Developing Economies (IDE). Japan External Trade Organization (JETRO). Chiba, Japan. http://www.ide.go.jp/English/Publish/Download/Asedp/077.html. Accessed on December 5, 2018.

Warner, Melanie. 1997. "How Tyson Ate Hudson in a Little Patch of Arkansas." *Fortune*, October 27. http://archive.fortune.com/magazines/fortune/fortune_archive/1997/10/27/233332/index.htm. Accessed on January 8, 2017.

Watson, R., and D. Pauly. 2001. "Systematic Distortions in World Fisheries Catch Trends." *Nature* 414:534–536.

Watts, D. C. H., Brian Ilbery, and Damian Maye. 2005. "Making Reconnections in Agro-Food Geography: Alternative Systems of Food Provision." *Progress in Human Geography* 29 (1): 22–40.

Weinberg, Carl. 2003. "Big Dixie Chicken Goes Global: Exports and the Rise of the North Georgia Poultry Industry." *Business and Economic History* 1:1–32.

Weis, Tony. 2010. "The Accelerating Biophysical Contradictions of Industrial Capitalist Agriculture." *Journal of Agrarian Change* 10 (3): 315–341. https://doi.org/10.1111/j.1471-0366.2010.00273.x. Accessed on January 2, 2019.

Weis, Tony. 2013a. *The Ecological Hoofprint: The Global Burden of Industrial Livestock*. London: Zed Books.

Weis, Tony. 2013b. "The Meat of the Global Food Crisis." *The Journal of Peasant Studies* 40 (1): 65–85. https://doi.org/10.1080/03066150.2012.752357. Accessed on January 10, 2019.

Weis, Tony. 2015. "Meatification and the Madness of the Doubling Narrative." *Canadian Food Studies/La Revue canadienne des études sur l'alimentation* 2 (2): 296–303.

Weiss, B. 2012. "Configuring the Authentic Value of Real Food: Farm-to-Fork, Snout-to-Tail, and Local Food Movements." *American Ethnologist* 39 (3): 614–626.

Welcomme, Robin L. 2011. "An Overview of Global Catch Statistics for Inland Fish." *ICES Journal of Marine Science* 68 (8): 1751–1756.

Wieger, Léon. 1927. *Chinese Characters: Their Origin, Etymology, History, Classification and Signification: A Thorough Study from Chinese Documents*. 2nd ed. New York: Catholic Mission Press.

Willer, H., and D. Schaak. 2016. "Organic Farming and Market Development in Europe." In *The World of Organic Agriculture: Statistics and Emerging Trends*, ed. H. Willer and J. Lernoud, 199–220. Bonn: Research Institute of Organic Agriculture, Frick and IFOAM–Organics International.

Winders, Bill. 2006. "'Sowing the Seeds of Their Own Destruction': Southern Planters, State Policy and the Market, 1933–1975." *Journal of Agrarian Change* 6 (2): 143–166.

Winders, Bill. 2009. *The Politics of Food Supply: U.S. Agricultural Policy in the World Economy*. New Haven: Yale University Press.

Winders, Bill. 2017. *Grains*. Malden, MA: Polity Press.

Winders, Bill, Alison Heslin, Gloria Ross, Hannah Weksler, and Seanna Berry. 2016. "Life after the Regime: Market Instability with the Fall of the U.S. Food Regime." *Agriculture and Human Values* 33 (1): 73–88.

References

Wittwer, Sylvan, Youtai Yu, Sun Han, and Lianzheng Wang. 1987. *Feeding a Billion: Frontiers of Chinese Agriculture*. East Lansing: Michigan State University Press.

Wong, Yin Khoon. 1990. *Unlocking the Chinese Heritage*. Singapore: Pagesetters Services Pte Ltd.

Woodruff, Judy. 2014. "Who's Behind the Chinese Takeover of World's Biggest Pork Producer?" *PBS NewsHour*. September 12, 2014. http://www.pbs.org/newshour/bb/whos-behind-chinese-takeover-worlds-biggest-pork-producer/. Accessed on February 14, 2017.

Woods, Clyde Adrian. 2017. *Development Arrested: The Blues and Plantation Power in the Mississippi Delta*. 2nd ed. New York: Verso.

Woosley, Michael, and Jianping Zhang. 2010. *China, People's Republic of: Livestock and Products Semi-Annual Report 2010*. United States Department of Agriculture Foreign Agricultural Service, GAIN Report No. CH10009. Beijing.

World Health Organization (WHO). 2015. "Vitamin A Fortification of Staple Foods." *E-Library of Evidence for Nutrition Actions (eLENA)*. World Health Organization. Last updated August 10, 2015. http://www.who.int/elena/titles/vitamina_fortification/en/. Accessed on December 4, 2018.

Wright, W., and S. L. Muzzatti. 2007. "Not in My Port: The 'Death Ship' of Sheep and Crimes of Agri-food Globalization." *Agriculture and Human Values* 24 (2): 133–145.

Wunder Sven. 2005. *Payments for Environmental Services: Some Nuts and Bolts*. Occasional Paper 42. Bogor, Indonesia: CIFOR.

Xiao Hong-b, Qiong Chen, Ji-min Wang, Les Oxley, and Heng-yun Ma. 2015. "The Puzzle of the Missing Meat: Food Away from Home and China's Meat Statistics." *Journal of Integrative Agriculture* 14 (6): 1033–1044.

Xiu, Q. Ma, Julia M. Verkuil, Helene C. Reinbach, and Lene Meinert. 2017. "Which Product Characteristics Are Preferred by Chinese Consumers When Choosing Pork? A Conjoint Analysis on Perceived Quality of Selected Pork Attributes." *Food Science and Nutrition* 5 (3): 770–775.

Yan, Hairong, Yiyuan Chen, and Ku Hok Bun. 2016. "China's Soybean Crisis: The Logic of Modernization and Its Discontents." *Journal of Peasant Studies* 43 (2): 373–395.

York, Richard, and Marcia Hill Gossard. 2004. "Cross-National Meat and Fish Consumption: Exploring the Effects of Modernization and Ecological Context." *Ecological Economics* 48 (3): 293–302.

York, Richard, Eugene A. Rosa, and Thomas Dietz. 2003a. "Footprints on the Earth: The Environmental Consequences of Modernity." *American Sociological Review* 68 (2): 279–300.

York, Richard, Eugene A. Rosa, and Thomas Dietz. 2003b. "STIRPAT, IPAT and ImPACT: Analytic Tools for Unpacking the Driving Forces of Environmental Impacts." *Ecological Economics* 46 (3): 351–365.

Zastiral, Sascha. 2014. "Half a Billion New Middle-Class Consumers from Rio to Shanghai." In *Meat Atlas: Facts and Figures about the Animals We Eat*, 48–49. Berlin; Brussels: Heinrich Böll Foundation and Friends of the Earth Europe. https://www.foeeurope.org/sites/default/files/publications/foee_hbf_meatatlas_jan2014.pdf. Accessed on December 19, 2018.

Zepeda-Millán, Chris. 2017. *Latino Mass Mobilization: Immigration, Racialization, and Activism*. Cambridge, UK: Cambridge University Press.

Zhang, Huijie, Shenggen Fan, and Keming Qian. 2005. "The Role of Agribusiness Firms in Agricultural Research: The Case of China." *Annual Meeting of the American Agricultural Economics Association*, Providence, RI, July 24–27. https://ideas.repec.org/p/ags/aaea05/19415.html. Accessed on January 10, 2019.

Zhang, Qian Forrest, and John Donaldson. 2008. "The Rise of Agrarian Capitalism with Chinese Characteristics: Agricultural Modernization, Agribusiness and Collective Land Rights." *China Journal* 60 (60): 25–47.

Zheng, P. 1984. *Livestock Breeds of China*. Report No. 46. Rome: FAO Animal Production and Health Series and China Academic Publishers.

Ziser, Michael, and Julie Sze. 2007. "Climate Change, Environmental Aesthetics, and Global Environmental Justice Cultural Studies." *Discourse* 29 (2): 384–410.

Zúñiga, Víctor, and Rubén Hernández-León. 2005. *New Destinations: Mexican Immigration in the United States*. New York: Russell Sage Foundation.

Contributors

Conner Bailey is Professor Emeritus of Rural Sociology at Auburn University. Prior to joining the Auburn faculty in 1985, Bailey worked at the International Center for Living Aquatics Resource Management (ICLARM, now the WorldFish Center) and in the Marine Policy Center at Woods Hole Oceanographic Institution. He began research on marine fisheries in the mid-1970s while working on his doctoral dissertation and on aquaculture in the early 1980s while working in Southeast Asia with ICLARM. He has taught tropical fisheries management at the Norwegian College of Fisheries, University of Tromsø (1998–2002) and has worked with various international agencies (FAO, World Bank, USAID) on issues related to fisheries management and aquacultural development.

Robert M. Chiles is an Assistant Professor in the Department of Agricultural Economics, Sociology, and Education and the Department of Food Science and a Research Associate in the Rock Ethics Institute at Penn State University. His scholarship examines the ethical and empirical dimensions of food system controversies, the interconnections between food culture and political economy, and the methods and social processes by which agricultural science is performed. Chiles's research has been published in *Sociologia Ruralis*, *Agriculture and Human Values*, *Journal of Agricultural and Environmental Ethics*, *Journal of Consumer Behaviour*, *Journal of Food Science Education*, and *Contexts* and in *Controversies in Science and Technology, Vol. 4: From Sustainability to Surveillance* (2014).

Celize Christy completed her MS in Rural Sociology and International Agricultural Development at Penn State University. With a background in animal science, her research interests lie at the intersection of livestock production, smallholder livelihoods, and Sub-Saharan African agricultural development. She has had experience in research projects and rural extension efforts with poultry farmers in the Kamuli District of Uganda. Her thesis examined the use of local knowledge and community networks on preventive poultry health measures in Rwanda.

Riva C. H. Denny is a Research Associate in the Department of Sociology at Michigan State University. Her research focuses on agriculture and the environment, as well as food systems and food security. She has a PhD in sociology from Michigan State University, and an MS in rural sociology from Auburn University. Her work has been published in *Rural Sociology* (2017 Best Paper Award winner), *Sociology of Development*, *Society and Natural Resources*, and *Environment Systems and Decisions*.

Carrie Freshour is an Assistant Professor in the Division of Social Sciences and History at Delta State University. Her research interests include the political economy of food work, social reproduction, and racialization in the U.S. South. She has conducted ethnographic fieldwork in Northeast Georgia, which included more than five months of working "on the line" in a large processing plant and conducting fifty oral histories with current and former poultry plant workers in order to understand the daily struggles and experiences of this racialized and gendered workforce.

Philip H. Howard is an Associate Professor in the Department of Community Sustainability at Michigan State University. He is the author of *Concentration and Power in the Food System: Who Controls What We Eat?* (2016), which won the Fred Buttel Outstanding Scholarly Achievement Award from the Rural Sociological Society in 2016. He is a former president of the Agriculture, Food and Human Values Society, and is currently a member of the International Panel of Experts on Sustainable Food Systems, and a member of the editorial board of *Agriculture and Human Values*. His visualizations of food industry changes have been featured in numerous media outlets.

Elizabeth Ransom is an Associate Professor in the School of International Affairs and a Senior Research Associate in the Rock Ethics Institute at Penn State University. She coedited *Rural America in a Globalizing World: Problems and Prospects for the 2010s* (2014). Her areas of expertise include international development and globalization, the sociology of agriculture and food, social studies of science and technology related to agriculture and food, and gender and development. For the past two decades one of her primary research interests has focused on livestock production in eastern and southern Africa. She has published articles in journals such as *Gender & Society*, *Journal of Rural Studies*, and *Rural Sociology*.

Thomas K. Rudel got his first exposure to tropical rain forests right after college when he worked as a Peace Corps volunteer in the Ecuadorian Amazon. It proved to be a pivotal experience! After earning a PhD in sociology, he carried out a series of studies of the social forces that have caused the recent destruction of so many tropical forests. He has written two books and more than twenty-five articles on cattle ranching and the loss of tropical forests. These studies have included both detailed, village-level studies in the Ecuadorian Amazon and quantitative analyses at the global scale. He is currently a Distinguished Professor of Human Ecology and Sociology at Rutgers University.

Mindi Schneider is an Assistant Professor of Sociology of Development and Change at Wageningen University in the Netherlands. She specializes in political economy of development, environmental sociology and political ecology, and international agriculture and rural development. Her current research centers on the social, political, and ecological transformations that accompany the industrialization of China's agro-food system, and on global food and agricultural politics more broadly. She has published in journals including *Geoforum*, the *Journal of Agrarian Change*, the *Journal of Peasant Studies*, and *Agriculture and Human Values*.

Nhuong Tran is a scientist at WorldFish based in Penang, Malaysia. He joined WorldFish in 2011 and since then has been working on various areas on fish supply and demand modeling, climate change impact and adaptation in aquaculture, and analysis of aquaculture system and value chain performance. Tran has published his research in journals such as the *American Journal of Agricultural Economics*, *World Development*, *Marine Policy*, *Marine Resource Economics*, *Agribusiness*, *Global Food Security*, and *Journal of Cleaner Production*.

Bill Winders is an Associate Professor in the School of History and Sociology at the Georgia Institute of Technology. He specializes in the areas of political sociology, social movements, the world economy, inequality, and the political economy of food and agriculture. His current research examines two issues: (1) the global meat industry, focusing on the production, consumption, and trade of meat in the world economy, and (2) food crises in the world economy, such as the 2008 food crisis that saw food prices and world hunger rise dramatically. He is the author of two books: *The Politics of Food Supply: U.S. Agricultural Policy in the World Economy* (2009), and *Grains* (2017).

Index

Page numbers followed by f refer to figures; page numbers followed by t refer to tables.

ADM, 1, 27, 90–91
Africa, sub-Saharan. *See also* Rwanda
  animal ethics in, 176–177
  aquaculture in, 57
  caloric intakes in, 172
  development strategies, 183–184
  dietary changes, 172
  GHG emissions, 146–147, 152–153, 158t, 161
  meat, demand for, 170, 172, 180
  nutritional transition in, 172–173
  South Africa, 122–123
  technological training needs, 184
Animal ethics, 142–143, 165n13, 177–178, 192
Animal rights
  activist movement, 167–168
  and development ethics, 169–170
  in development programs, 173–174, 179–180
  versus food security, 169
  recognizing, importance of, 184
  welfare, 178, 195
Antibiotic resistance, 94–95
AquaBounty Technologies (ABT), 71
Aquaculture
  in Africa, 57
  versus animal agriculture, 65–66
  in Asia, 57–58, 74n1
  in Canada, 61, 71
  certification initiatives, 73
  CFAO use, 56
  in Chile, 57, 61
  corporate concentration in, 56–57, 65–71, 73, 186
  and disease, 57–58, 59, 60–61, 63, 65, 66, 69–70
  and domestic consumption, 64
  environmental impacts, 59–62
  extensive versus intensive systems, 58–59
  feed, 57–58, 59–60, 68–69, 72
  future of, 71–74
  genetic engineering, 71
  growth and scale of, 55, 57
  history of, 57–58
  selective breeding, 69–70
  small-scale producers, 72, 74
  societal impacts, 62–65
Aquaculture Stewardship Council (ASC), 197, 198
Archer Daniels Midland (ADM), 1, 27, 90–91
Argentina
  anti-GE activism in, 29
  beef exports, 103, 107, 110

Argentina (cont.)
  corn industry, 26–27
  geographic advantages, 37
  GE seed use, 15, 28
  soybean industry, 26–27
Asia. *See also* China; India
  aquaculture in, 57–58, 74n1
  GHG emissions, 146–147, 150, 158t
  Indonesia, 152
  Japan, 5, 11–12, 67–68
  meat imports, 11–12
  pork EI, decline in, 163
  South Korea, 5, 11t, 12
  Thailand, 64
  vegetarianism in, 190
  Vietnam, 12, 22n3, 58, 64, 74n2
Australia
  geographic advantages, 37
  GHG emissions, 158t
  live animal transport practices, 143, 178
  meat companies in, 49f
  meat consumption, 4, 97n5
  meat exports, 10t, 11, 12, 77

BASF, 27
Bayer, 27–28
Beef, 4, 5f, 37, 41, 122, 172
Beef exports
  American, 10–11
  Argentinian, 103, 107, 110
  Brazilian, 10–11, 51, 102–103, 117
  Indian, 10t, 11, 12, 22n3, 75–76
  global, 10–11, 13
Beef imports, 11–13, 107, 110, 116
Beef industry. *See also* Ecuadorian cattle ranching
  Chinese, 81f
  corporate concentration in, 15, 31, 50
  Ecuadorian, 107, 110–111, 116, 117, 188
  GHG emissions, 147–150, 152, 157–161, 163

Indian, 22n3
JBS in, 46, 48–50
production increases, global, 6–8
sustainability initiatives, 51, 198
trade in, 10–13, 37
Tyson in, 36–37, 38
Beer industry, 52
Belton, Ben, 64
Bentsen, H. B., 69
Beyond Meat, 191
Bichler, Shimshon, 34
Brasil Foods (BRF), 47
Brazil
  China, trade with, 90
  corn production, 26
  deforestation in, 162
  feed costs in, 50
  GE seed planting, 28
  GHG emissions, 150, 161
  meat consumption, 51
  national champions development strategy, 46
  soybean production, 26
  subsidies, finance, 46–47, 50
Brazilian exports
  beef, 10–11, 102–103
  commodities generally, 51
  corn, 14, 27, 50
  pork, 10–11, 102–103
  poultry, 10
  soybean, 14, 27, 36, 50, 90
Bunge, 27, 90–91
Busch, Lawrence, 177
Bush, Faye, 125, 126
Bush, Simon R., 64

CAFOs. *See* Confined animal feeding operations
Canada
  anti-GE activism, 29
  aquaculture in, 61, 71
  farm size trends, 28
  GE seed use, 15

Index

meat consumption, 4
meat exports, 10t, 11, 12
Canola, 25, 28, 29
Carcass yields, 159–160, 161–162, 164n10
Cargill
  acquisition of EWOS, 69
  animal welfare policies, 195
  Chinese holdings, 123
  and feed industry, 27
  and GRSB, 198
  pork business acquisition by JBS, 35f, 48f, 50
  and soybean industry, 90–91
Carp, 59, 62, 68, 70
Catfish, 57, 58, 59, 61, 69, 70, 74n2
Cattle ranching. *See* Ecuadorian cattle ranching
Chicken. *See* Poultry
Chile, 57, 61
China
  agricultural modernization, 79, 86–88
  aquaculture in, 58, 74n1
  beef production, 81f
  Brazil, trade with, 90, 122
  corn imports, 43
  corn production, 26
  corporate concentration in, 44–45, 86, 88
  dietary guidelines, 95
  diets in, 93–94
  dragon head enterprises, 45, 87–88, 92–93
  environmental pollution, 95
  food networks, non-industrial, 97
  food safety concerns, 94, 96–97, 100n26
  GHG emissions, 150, 151–152, 161
  Golden Pig year, 82–83, 98n10
  Household Responsibility System (HRS), 79
  liberalization, economic, 79
  meat production, 81t
  meat trade, 10t, 11t, 12
  obesity rates, 94
  pigs in, 82–84
  pork in, 43, 46, 80–84, 93
  poultry in, 80, 81f, 122
  reform era, 79, 83–84, 86–87
  and Round Table on Responsible Soy, 46
  seafood exports, 67
  smallholder production, 187
  soybean imports, 36, 43, 79, 89–91
  soybean production, 26, 90–91
  subsidies, 41–43, 84–87, 98n13
Chinese meat consumption
  and Cultural Revolution, 83–84
  health impacts, 94
  historical, 82
  increases in, 5, 12, 46, 80, 93, 97n5, 98n6
  pork, 46, 80–82, 84, 89, 92, 93
  reduction, calls for, 96
  and social status, 83
Chinese pork industry. *See also* WH Group Limited (China)
  boom in, 79, 86, 95–96, 163n4
  breeds used in, 89, 99n21
  CAFOs and, 85–86, 89, 91, 94, 95, 186–187
  corporate concentration in, 85, 88, 92–93
  development of, 84–91
  downturn in, 96
  dragon head enterprises in, 87–88, 92–93
  effects of, negative, 91–92, 94, 187
  environmental impacts, 94–95, 100n28
  feed and soybeans in, 89–90
  food safety in, 94–95, 96–97
  future directions in, 96–97
  as global, 81
  health concerns, 94
  imports, 43

Chinese pork industry. (cont.)
  industrialization of, 84, 86–89
  offshoring, 90–91
  peasant production, 92
  production and emissions, 151
  size of, 79–80, 97n4
  smallholders in, 88
  subsidies, 43, 84–86, 98n13
Climate change, 142, 162, 165n14, 198–199. *See also* Greenhouse gas (GHG) emissions
Cobb-Vantress, 38–39
Confined animal feeding operations (CAFOs), 7
  in aquaculture, 56
  and Chinese pork industry, 85–86, 89, 91, 94, 95, 186–187
  and climate change, 162, 199
  feed used, 25
  growth of, 92
  labor concerns, 17
Corn industry, global, 13, 22n5, 25–27
Corn trade, 14, 27, 43, 50
Corporate concentration, 2, 31, 56, 186. *See also* Vertical integration
  American, 15
  aquaculture industry, 56–57, 65–71, 73, 186
  beef industry, 15, 31, 50
  in China, 44–45, 85, 88, 92–93
  concentric, 31
  cultured meat industry, 196
  depth versus breadth, 34
  and distant places, 34, 36
  effects of, 32
  farming industry, 28–29
  feed industry, 27–28
  GE seed industry, 15
  horizontal integration, 31, 93, 122
  JBS acquisitions, 48–50
  meat processors, 15, 35f
  poultry industry, 122
  reasons for, 15
  resistance to, 52–53
  seafood industry, 64, 71–72
  seed industry, 14–15, 27–28
  Tyson acquisitions, 38, 40–41f, 51
  WH Group acquisitions, 42–43, 44–45f
Corporate concentration, subsidies enabling, 32
  competitive advantages, 33
  feed, 37, 38, 39, 50
  finance, 41, 42, 46–47, 51
Cultured meat, 195–196

Denmark, 5, 22n2
Development programs, 173–174
Development strategies, 183–184
Dietary substitution effects, 172
Dietary trends, global, 172
Dietz, Thomas, 172
Diseases
  antibiotic-resistant, 94
  in aquaculture, 57–58, 59, 60–61, 63, 65, 66, 69–70
  avian influenza, 38
  in beef industry, 103
  and climate change, 199
  diet-related, 93–94, 173, 187
  and genetic uniformity, 36, 38, 52
  and intensive animal raising, 149
  in pork industry, 84
Dow Chemical Company, 27–28
DuPont, 15, 27–28, 90

Ecuador
  beef imports, Argentine, 107, 110, 116
  cattle rancher surveys, 107–116
  economic crisis of 1998, 106–107, 110
  land ownership patterns, 104–106
  Macas, 106, 108
  Morona Santiago region, 104–106, 118
Ecuadorian cattle ranching
  and carbon offsets, 114

Index

cattle importation, 111
cattle tethering, 119n2
economic crisis of 1998, 107, 110
and globalization, 103, 117–118
herd differentiation, 111
herd sizes, 108–111
household dependence on, 108
land cleaning, 112
landholding sizes, 108–110
land rentals, 106, 111–112
owning versus renting land, 112–116
and road access, 108
silvopastures, 107, 113, 114, 116
small-scale, 104–106
tree seedlings in, 113–114
Ecuadorian mestizos people
financing, access to, 106, 111
herd restocking advantages, 111
herd sizes, 109–110, 111
landholding sizes, 109
land ownership, 104–106
land renting, 111–112
migration rates, 109
silvopastures, 115
and sustainability, 115
Ecuadorian Shuar people
crops grown, 108
financing, access to, 106, 111
herd sizes, 109–110, 111
landholding sizes, 109
land ownership, 104–106
land renting, 111–112
migration rates, 109
natural-resource-degrading poverty traps, 115
silvopastures, 115
Emancipatory empiricism, 141–142
Emissions intensity (EI), 147, 149, 153, 160, 164n10. *See also* Greenhouse gas emissions intensity statistical analysis
English, Philip, 175, 182
Errington, Frederick, 77

Escapes, aquaculture, 61–62
European Union
animal welfare policies, 195
corn production, 26
GE seeds in, 23n7
GHG emissions, 150–152, 158t, 161
meat consumption rates, 5–6
meat processors in, 51
meat trade, 10t, 11t, 12
organic foodstuff sales, 192
vegetarianism in, 190
Expansion effect, dietary, 172
Externalities, 34

Farmers
and activism, 29
and animal ethics, 184, 192
aquaculture, 63, 69
Chinese, 82, 83, 88, 96–97, 99n21
and cultured meat, 196
decline in, 28, 32, 43, 96, 196
Ecuadorian, 103, 113–114
and feed industry, 27, 29
Georgian, 124–125
and GE seeds, 26, 28, 29
Mexican, 131, 139n12
Rwandan, 176–177, 179, 182
subsidies for, 37
and technology transparency, 184
and Tyson, 37, 39, 131
Feed conversion, 37, 42, 60, 85, 161
Feed industry
aquaculture, 57–58, 59–60, 68–69, 72
Brazilian, 26, 50
Cargill and, 27
and climate change, 162, 165n14
corn in, 13, 14, 22n5, 25–27, 43, 50
and farmers, 27, 29
and food insecurity, 16
land use, 13
North, global, 29
production increases, 25
and seafood industry, 59, 68–69

Feed industry (cont.)
  South, global, 29
  soybeans in, 25, 89–90
  subsidies, 25, 34, 37, 38, 39, 50
  trade in, 27
Feed industry, growth of
  and corporate concentration, 27–28
  factors enabling, 13–14, 22n5, 25–28
  and farm consolidation, 28–29
  and GE seeds, 13–16, 28–29
  and liberalization, 14, 27
  and meat industry growth, 25, 29
  statistics on, 13, 26
Fish. *See* Seafood
Flexitarianism, 191
Food and Agriculture Organization (FAO), 17, 66, 68, 74n1, 97n1, 149, 161, 163n1, 164n6, 199
Food fortification, 180–181
Food security
  versus animal rights, 169
  aquaculture increasing, 64–65
  contemporary statistics on, 75, 78n1
  declining, 16–17, 75, 186–187
  and feed industry, 16
  in India, 75–76
  and labor issues, 187
  and meat industry, 16
  and meat industry growth, 75–77
  and resource conflicts, 76
  in Rwanda, 175, 180–183
  and seafood industry, 64–65
  and trade, 75–76
Food sovereignty, 17, 197
Forkman, Björn, 178
France, 5–6, 29
Free trade agreements, 14, 16, 27, 129–131
Freudenberg, William, 36

Genetically engineered (GE) seeds
  activism against, 29
  in Argentina, 15, 28
  in Brazil, 28
  in Canada, 15
  and corporate concentration, 15, 25–26
  corporate control, enabling, 28
  countries using, 15, 23n7
  and farmers, 26, 28, 29
  and feed industry growth, 13–16, 28–29
  in the United States, 28
Georgia, 124–125, 130, 133–134, 139n19
Germany, 6
Gewertz, Deborah, 77
Ggombe, Kasim, 175, 182
Ghent, Belgium, 191
Gjøen, M. H., 69
Global Roundtable on Sustainable Beef (GRSB), 198
Gossard, Marcia Hill, 172
Government subsidies. *See* Subsidies, government
Greenhouse gas (GHG) emissions, 146, 163n1
  from beef, 147–150, 152, 157–161, 163
  and carcass yield, 159–160, 161–162, 164n10
  from enteric fermentation (ruminants), 146, 147, 163nn1–2
  FAO calculation method, 149
  and feed production, 163n3
  increases in, globally, 146–147, 161, 163
  from manure, 146, 147, 159
  North America, 150–151, 158t
  from pork, 147–149, 151, 158t, 159
  from poultry, 147–149, 151–152, 158t, 159, 164n6
  and social-environmental justice, 145–146
  and technology, 161
  versus total GHG emissions, 145

Greenhouse gas emissions intensity, 147, 149, 153, 160, 163n4, 164n10
Greenhouse gas emissions intensity statistical analysis
approach and design, 153–156, 164nn8–9
limitations of, 162–163, 165n12
results and discussion, 154, 157–161, 189
Green Revolution, the, 65

Heffernan, William D., 15
Heifer International, 173–174
Horizontal integration, 31, 93, 122

India
beef exports, 10t, 11, 12, 22n3, 75–76
food insecurity, 75
income increases, 194
meat industry, 194
vegetarianism in, 190, 194, 200n1
Indonesia, 152
Industrialization, political economy of, 86
Industrial model of animal agriculture, 37
Intensive production, 3. *See also* Confined animal feeding operations
and animal ethics, 142–143, 189
in aquaculture, 56–64, 66, 69, 70, 73, 187
effects of, 3, 91
and GHG emissions, 149, 154, 161, 162, 163
International Livestock Research Institute (ILRI), 173
Iran, 152
Iraq, 11t, 12

Japan, 5, 11–12, 67–68
JBS (Brazil), 31, 46–51

Jewel, Jesse, 125
Jinluo (China), 93

Kirwan, James, 168

Lab-grown meat. *See* Cultured meat
Labor in American poultry industry
black, 126–132, 134, 137
discipline problems, 126–127
disputes, 126
hiring practices, 125–126
immigrants, undocumented, 124, 130–134, 137
Immigration Customs Enforcement (ICE) raids, 132–133
Jewell's influence on, 125
and Jim Crow, 125
Latina/o, 130–132
and line speed increases, 135–136, 137–138
organizing and unions, 127–130, 132, 134, 137
power, loss of, 129–130
precarity of, 134–135, 137
women, 125–129, 134, 137
workforce increases, 129
Labor exploitation, American, 76–77
Labor movements, need for, 138
Land grabs, 76
Land usage
and discount rates, 101, 113
environmental stewardship of, 102
owners versus renters, 101–102, 112–116
sustainable, 103, 116
Lassen, Jesper, 178
Latin America, 57, 61, 65, 102–103. *See also* Argentina; Brazil; Chile; Ecuador; Mexico; South America
La Via Campesina, 196–197
Lazzarini, Sergio, 47
Liberalization, 14, 27, 79. *See also* Free trade agreements

Lifestyle politics, 191
Little, Jo, 64
Live animal transport, 143
Local food, 168, 171, 182, 191–192
Louis Dreyfus, 27, 90–91

Mangroves, 62–63
Marfrig (Brazil), 47, 50
Marine Stewardship Council (MSC), 197–198
Masiga, W. N., 176–177
McSharry, Patrick, 175, 182
Meat consumption. *See also* Chinese meat consumption; Pork consumption
 and affluence, 77, 172
 American, 4–6
 Australian, 4, 97n5
 beef, 4, 5f, 51, 122
 Brazilian, 51
 Canadian, 4
 declines in, 4–6, 191
 in the European Union, 5–6
 factors influencing, 6, 172
 increases in, 3, 4–6, 12
 and masculinity, 200n2
 Mexican, 12
 North, global, 4–6, 191
 poultry, 4, 122
 Russian, 12
 ubiquity of, 145
Meatification, 46, 52, 171–172
Meat industry, 1, 4–6, 8–9. *See also* Beef industry; Pork industry; Poultry industry
 effects of, 7, 16–17, 91–92, 162
 organic production in, 194
Meat industry, growth of
 causes of, 1, 173
 corporate-state partnerships, 185–186
 economic concerns, 16–17
 effects of, 1, 3, 16–17, 77
 environmental concerns, 17
 and external inputs, 36
 factors enabling, 7, 13–16, 29
 and the feed industry, 25, 29
 and food insecurity, 75–77
 governments enabling, 2
 in India, 194
 and labor exploitation, 76–77
 and land grabs, 76
 and population growth, 2–3
 production and slaughter rates, 3, 6–8
 and trade, 75–76
 value increases, 2
Meat industry, solutions for, 189, 190t
 animal welfare policies, 195
 consumer-oriented, 189–194
 contexts, socio-political, 194
 cultured meat, 195–196
 food sovereignty, 197
 organic production, 194
 organizational responses, 195–198
 smallitics, 193–194
 sustainability initiatives, 197–198
 veganism, 168–169, 191, 193
 vegetarianism, 190–191, 193–194, 200n1
Meatless Mondays, 191, 193
Meat replacements, 191, 195–196
Mexico
 agricultural exports, 131
 corporate concentration in, 43, 48f, 50
 meat consumption, 12
 meat imports, 11t, 12
 poultry production, 122
Monsanto, 1, 15, 27–29, 90
Morris, Carol, 168
Multi-stakeholder initiatives (MSIs), 197–198
Munyua, S. J. M., 176–177

National Beef Packing, 50
Neoliberal diet, the, 4, 77
Netherlands, 6

Index

Netting, Robert, 102, 114, 116
Nitzan, Jonathan, 34
North, global
  and aquaculture investments, 72, 73
  colonialism of, 169, 183
  diets, plant-based, 180
  feed industry, 29
  meat consumption rates, 4–6, 191
  meat exports, 10–11, 12
  meat industry value, 8
  meat processor growth, internal, 34
  obesity-hunger paradox, 77
  seafood imports, 57, 59, 64–65, 67–68
  trade and power, 12–13
North America, 10–11, 150–151, 158t, 192. *See also* Canada; United States of America
North American Free Trade Agreement (NAFTA), 27, 129–131
Norway, 57, 61, 67, 68–69

Obesity-hunger paradox, 77, 187
Oligopolies, 31
Oliveria, Gustavo, 90
Organic foodstuffs, 192, 194

Pangasius catfish, 58–59
Patel, Raj, 122
Pigs, role in China, 82–84
Pilgrim, Bo, 49–50
Pilgrim's Pride, 49–50, 133, 139n11
Ponte, S., 197–198
Population, global, 2–3, 6
Pork
  in China, 43, 80–84, 93
  feed conversion rate, 37
  production, global, 6–8
  rise of, 79
  trade of, 10–12, 43, 102–103
Pork consumption
  Chinese, 46, 80–82, 84, 89, 92, 93
  Danish, 22n2
  rates, global, 4, 122

Pork industry. *See also* Chinese pork industry
  corporate consolidation in, 35f, 48f, 50
  diseases in, 84
  GHG emissions, 147–149, 151, 158t, 159
Poultry, 4, 10–12, 37–38, 122
Poultry industry
  animal welfare in, 165n13, 178–179
  Chinese, 81f, 122, 123
  corporate concentration in, 38, 122
  GHG emissions, 147–149, 151–152, 158t, 159, 164n6
  production, global, 6–8
  in the South, global, 122–123
  South African, 122–123
Poultry industry, American. *See also* Labor in American poultry industry; Tyson Foods Inc.
  Acme Chicken Processing (ACP), 130–135, 138n1
  and Agricultural Adjustment Act (AAA), 124
  class struggles in, 124–135, 137
  and cotton industry, 124–125
  disability, 135, 136, 139n20
  and Economic Justice Coalition (EJC), 132
  and Food Safety Inspection Services (FSIS), 135–136
  in Georgia, 123–124
  globalization, enabling, 131, 135
  growth of, 129
  HIMP, 136
  line speeds, 123, 130, 135–136, 137–138, 140n22
  and Mexico, 131
  and NAFTA, 129–130, 131
  National Chicken Council (NCC), 136, 137
  origins of, 124–125

Poultry industry, American. (cont.)
    size of, 121
    subsidies, financial, 125

Resource conflicts, 62–63, 76
Romania, 43
Rosa, Eugene A., 172
Round Table on Responsible Soy, 46, 51
Ruminants, 146, 147, 163n1
Russia, 11t, 12, 23n6, 46
Rwanda, 170–171, 174
    agricultural sector changes, 176
    agriculture, non-animal, 181–182
    and animal welfare standards, 176–180, 189
    changes in, contemporary, 174–175
    development projects, 179
    development strategies, 175–176, 182
    diets, traditional, 180
    food fortification in, 180–181
    food security in, 175, 180–183
    ICT industry, 182–183
    intercropping, 182
    livestock industry, 170, 175, 177
    meat, demand for, 180
    MINAGRI, 175–176
    poultry industry, 170–171, 175–176
    service industry, 182
    smallholder farms, 176, 177

Salmon, 57, 59, 61, 68–70
Sandøe, Peter, 178
Saudi Arabia, 12
Seafood. *See also* Aquaculture
    carp, 59, 62, 68, 70
    catfish, 57, 58, 59, 61, 69, 70, 74n2
    Chinese exports, 67
    demand for, 55–56
    production of, 55
    salmon, 57, 59, 61, 68–70
    shrimp, 56, 58–59, 61–63, 68, 70
    supply of, 55

tilapia, 57, 59, 68, 69, 70
trade of, 55
Seafood industry. *See also* Aquaculture
    corporate concentration in, 64, 71–72
    feed production, 68–69
    fishing vessel statistics, 66
    and food security, 64–65
    genetic engineering in, 71, 72
    growth of, 64
    Japanese corporations, 68
    Thai, 64
    trade, international, 64–65, 67–68
Seed industry, 14–15, 27–28, 90.
    See also Genetically engineered (GE) seeds
Shrimp, 56, 58–59, 61–63, 68, 70
Shuanghui. *See* WH Group Limited (China)
Silvopastures, 107, 113–114, 116–117
Singer, Peter, 167
Singer, R., 193
Smallholders
    Chinese, 88, 187
    decline in, 17
    in food sovereignty movement, 197
    land access, 3
    livestock dependence, 187
    Rwandan, 176, 177
    and sustainability initiatives, 198
Smallitics, 193–194
Smil, Vaclav, 172
Smithfield, 42, 43, 46. *See also* WH Group Limited (China)
Solutions. *See* Meat industry, solutions for
South, global
    and animal rights, 169, 179–180
    and climate change, 199
    feed industry, 29
    meat, rising demand for, 171, 173
    meat imports, 12
    poultry industry, 122–123

Index

seafood exports, 64–65, 67–68
Tyson in, 122
South Africa, 122–123
South America. *See also* Argentina; Brazil; Ecuador
Chile, 57, 61
GHG emissions, 146–147, 150–151, 158t
meat exports, 10–11, 12
soybean production, 26
South Korea, 5, 11t, 12
Soybean industry
American, 26, 90
Argentinian, 26–27
Brazilian, 26
Cargill and, 90–91
Chinese, 26, 90–91
growth of, 13, 22n5, 26
land use, 26
Louis Dreyfus and, 90–91
seeds, control of, 90
South American, 26
Soybeans in feed, 25, 89–90
Soybean trade
American, 14, 27, 36, 90
Brazilian, 14, 27, 36, 50, 90
Chinese, 36, 43, 79, 89–91
Speciesism, 167–168
STIRPAT, 153, 164
Subsidies, government. *See also* Corporate concentration, subsidies enabling
acquisition financing, 34
American, 37, 38–39, 43, 125
and biophysical barriers, 33–34
Brazilian, 46–47, 50
Chinese, 41–43, 51, 84–87, 98n13
dominant firms benefiting, 33–34
effects of, negative, 34, 36, 52
for farmers, 37
of feed, 13, 25, 34
feed industry, 25, 34, 37–39, 50
and globalization, 33

JBS, 46–47, 50–51
legitimation-accumulation tensions, 32, 33
and market concentration, 32, 52
and meatification, 52
rationalizatons for, 36, 45–46
and social barriers, 34
Sustainability initiatives, 197–198
Syngenta, 15, 27, 90

Thailand, 64
Tilapia, 57, 59, 68, 69, 70
Trade, 9–13
free trade agreements, 14, 16, 27, 129–131
geographic dimensions, 12
increases in, 75
live animal, 143
power in, 12–13
soybean, 27, 90–91
Trade, exports. *See also* Beef exports; Brazilian exports; United States of America, exports
agricultural, Mexican, 131
meats, global (non-seafood), 10–12, 77
seafood, 64–65, 67–68
Trade, imports
beef, global, 11–13, 107, 110, 116
cattle, Ecuadorian, 111
corn, Chinese, 43
meats, global (non-seafood), 11–13
meats, Russian, 23n6
pork, Chinese, 43
poultry, South African, 122–123
seafood, global, 57, 59, 64–65, 67–68
South, global, 12
soybean, Chinese, 36, 43, 79, 89–91
Transnational corporations (TNCs), 27–28. *See also* Corporate concentration
Tyson Foods Inc. (United States)
acquisitions, 38, 40–41f, 51

Index

oods Inc. (United States) (cont.)
 al welfare policies, 195
 b-Vantress subsidiary, 38–39
 porate motto, 38
 visions acquired by JBS, 50
 mmigrant labor, undocumented, 139n11
 labor exploitation, 39, 76
 in Latin America, 50
 lobbying efforts, 39–40
 meat replacement investments, 191
 in Mexico, 131
 public relations, 40
 size of, 31, 36–37, 68, 139n21
 in the South, global, 122
 strength of, contemporary, 51–52
 subsidies received, 37, 38, 39

Unions, 128–130, 138
United States of America
 animal deaths and weather, 199
 aquaculture history, 57
 corn production, 26
 corporate concentration in, 15
 farm size trends, 28
 GE seed planting, 28
 GHG emissions, 150–152, 161
 labor exploitation in, 76–77
 meat consumption, 4–6
 meat imports, 11t, 12
 organic foodstuff sales, 192–193
 seafood imports, 67
 soybean production, 26, 90
 vegetarianism in, 190
United States of America, exports of
 corn, 14, 27
 meats, all, 10–11
 poultry, 122–123
 soybean, 14, 27, 36, 90

Veganism, 168–169, 191, 193
Vegetarianism, 190–191, 193–194, 200n1

Vertical integration
 in animal agriculture generally, 56, 122
 in China, 45, 86, 88
 meat processing industry, 31
 in seafood industry, 64, 65, 67–69, 74
 of Tyson, 37
 of WH Group, 42–43
Vietnam, 12, 22n3, 58, 64, 74n2

Weis, Tony, 46
WH Group Limited (China), 31, 41–43, 44–45f, 51, 93
World Organization for Animal Health (OIE), 195
World Trade Organization (WTO), 14, 27

York, Richard, 172
Yurun (China), 93

Food, Health, and the Environment

Series Editor: Robert Gottlieb, Henry R. Luce Professor of Urban and Environmental Policy, Occidental College

Keith Douglass Warner, *Agroecology in Action: Extending Alternative Agriculture through Social Networks*

Christopher M. Bacon, V. Ernesto Méndez, Stephen R. Gliessman, David Goodman, and Jonathan A. Fox, eds., *Confronting the Coffee Crisis: Fair Trade, Sustainable Livelihoods and Ecosystems in Mexico and Central America*

Thomas A. Lyson, G. W. Stevenson, and Rick Welsh, eds., *Food and the Mid-Level Farm: Renewing an Agriculture of the Middle*

Jennifer Clapp and Doris Fuchs, eds., *Corporate Power in Global Agrifood Governance*

Robert Gottlieb and Anupama Joshi, *Food Justice*

Jill Lindsey Harrison, *Pesticide Drift and the Pursuit of Environmental Justice*

Alison Alkon and Julian Agyeman, eds., *Cultivating Food Justice: Race, Class, and Sustainability*

Abby Kinchy, *Seeds, Science, and Struggle: The Global Politics of Transgenic Crops*

Vaclav Smil and Kazuhiko Kobayashi, *Japan's Dietary Transition and Its Impacts*

Sally K. Fairfax, Louise Nelson Dyble, Greig Tor Guthey, Lauren Gwin, Monica Moore, and Jennifer Sokolove, *California Cuisine and Just Food*

Brian K. Obach, *Organic Struggle: The Movement for Sustainable Agriculture in the United States*

Andrew Fisher, *Big Hunger: The Unholy Alliance between Corporate America and Anti-Hunger Groups*

Julian Agyeman, Caitlin Matthews, and Hannah Sobel, eds., *Food Trucks, Cultural Identity, and Social Justice: From Loncheras to Lobsta Love*

Sheldon Krimsky, *GMOs Decoded: A Skeptic's View of Genetically Modified Foods*

Rebecca de Souza, *Feeding the Other: Whiteness, Privilege, and Neoliberal Stigma in Food Pantries*

Bill Winders and Elizabeth Ransom, eds., *Global Meat: Social and Environmental Consequences of the Expanding Meat Industry*